PRACTICAL MANUAL OF PLANT ECOLOGY AND PLANT PHYSIOLOGY

PRACTICAL MANUAL OF PLANT ECOLOGY AND PLANT PHYSIOLOGY

By

Dr. S. Sundara Rajan

Prof. and Head, Dept. of Botany
St. Joseph's College
Bangalore - 560 025

ANMOL PUBLICATIONS PVT. LTD.

NEW DELHI - 110 002 (INDIA)

ANMOL PUBLICATIONS PVT. LTD.
4374/4B, Ansari Road, Daryaganj
New Delhi - 110 002
www.anmolbooks.com

Practical Manual of Plant Ecology and Plant Physiology

First Edition, 2001

ISBN 81-261-0963-7

PRINTED IN INDIA

Published by J.L. Kumar for Anmol Publications Pvt. Ltd., New Delhi - 110 002 and Printed at Tarun Offset Press, Delhi.

Contents

Part One - Plant Ecology

Part Two - Plant Physiology

Preface

Practical Manual of Plant Ecology and Plant Physiology is a text book meant for B.Sc., B.Sc (Hons.) and M.Sc., students taking botany as a subject. It is also useful for students of Agriculture and Microbiology.

The manual is easy to use and guides the students step by step in conducting experiments in the laboratory. The text material is illustrated with a number of neatly drawn diagrams. The experiments chosen are such that they can be used by students of all Universities.

Constructive criticisms are welcome from the users of the book.

I am thankful to ***Mr. J.L. Kumar,*** M.D. **Anmol Pulications** for publishing this book.

Dr. S. Sundara Rajan
Prof. and Head

Dept. of Botany
St. Joseph College,
Bangalore - 560 025.

Part - One
Plant Ecology

1

STUDY OF VEGETATION

Vegetation study or study of the community structure is essential for understanding the dynamics of a plant community. Many of the features of a plant community such as density, species frequency etc are best expressed in terms of quantitative parameters. There are three methods of field study of the analytical characters. This are quadrat method, line transect method and point frame method. The following are some of the field experiments pertaining to the above.

1. To determine the Minimum size of the quadrat by species Area-curve method.

Requirements : Scale, four nails, A few meterres long string, hammer, graph sheet, herbarium sheet, cellotape etc.

Procedure: Hammer 3 nails to the ground in a 'L' shape and tie a piece of string. Now with another nail and a piece of string fashion a square of 10 x 10 cm. Make a note of the various species present in the area. This squarish area is called a quadrat. Increase the size of the quadrat to 20 x 20 cm^2 and note the species present. In the same way go on increasing the quadarat size to 30 x 30 Cm^2, 40 x 40 Cm^2 etc each time recording the number of species. Continue to increase the size of quadrat as long as there is increase in the number of species falling in the area. Stop increasing the size of the quadrat when there is no corresponding increase in the number of species.

Record the observations on a graph sheet. Size of the area to be marked on the X asis and the corresponding increase in the number of species on 'Y' axis. When you join the points on the sheet it would result in a curve. This is called the 'species area curve'. You will notice that initially there will be geometric increase in the number of species with the corresponding increase in the size of quadrat.

Note the point where the curve starts flattening. Draw straight lines from this point to Y and X axis. The square area that will be obtained is the minimum quadrat size.

Conclusion : Species area curve is used to obtain the minimum size of quadrat while sampling the species composition of a plant community. Number of species (not individuals) in an area is a can be used as a parameter of diversity the number of quadrats. The curve obtained will indicate (where it starts to flatten) the point from which the minimum number of quadrats required can be found out.

2. *To determine the minimum number of quadrats to be laid down in the field.*

Requirements : Quadrat of known size, graph sheet, herbarium sheet, cellotape etc.

Procedure : Lay down about 30-50 quadrats of known size (as decided in the previous experiment) randomly at different locations in the area of study. List the various species found in each quadrat and find out the accumulation of species in each quadrat as shown in the table.

Species	1	2	3	4	5	6	7	8	9	10
A	+	+	-	-	+	-	+	-	+	-
B	-	-	+	-	+	-	-	+	-	+
C	-	+	-	+	-	+	-	-	+	+
D	+	-	-	-	-	-	+	+	+	+
E	-	-	+	+	+	+	+	+	+	+
F	-	-	-	-	-	+	-	-	-	-
G	+	+	+	+	+	+	+	+	+	+
H	-	-	-	-	-	-	-	-	-	+
I	+	-	-	+	-	+	+	-	+	+
J	+	+	+	-	-	-	+	-	+	-
Accumulation of total No. of species	5	4	4	4	4	5	6	4	7	7

Take a graph sheet and plot the accumulation of number of species (on the Y axis) against the numbe of quadrats. The curve obtained will indicate (where it starts to flatten) the point from which the minimum of quadrats required can be found out.

3. To determine the density, frequency and abundance of different species in an area

Requirement : Quadrat of known size, graph sheet, herbarium sheet, cellotape, pencil etc.

Procedure : Lay down a known number of quadrats (as determined in experiment 2) of definite size (as determined in experiment 1) in the area of study. List the various species present in each quadrat; also count the number of individual plants in each species (in every quadrat). Tabulate the data as shown in the following table.

Name of Species	No. of Quadrats land down 1234 etc	Total No. of individuals in each species studied	No. of Quadrats in which species occur	Total number of quadrats	Comments
A					
B	*2*4 -- --	6	2	4	
C					
D					

* No of individual plants occuring in that quadrat. From the data tabulated as above calculate the following.

Density : This is defined as the number of individual plants of a given species in a unit area.

$$\text{Density} = \frac{\text{Total number of individuals of the species}}{\text{Total number of quadrats studies.}}$$

$$\text{or} = \frac{\text{Toal number of quadrats in which the species occurs}}{\text{Total number of quadrats studied}} \times 100$$

Frequency : This is expressed in terms of percentage of distribution of species in a given area.

$$\text{Frequency (\%)} = \frac{\text{Total number of quadrats in which species is present}}{\text{Total number of quadrats studied}} \times 100$$

Abundance : This refers to the total number of individuals of a species in a given area.

$$\text{Abundance} = \frac{\text{Total number of individuals in all the quadrats}}{\text{Total number of quadrats in which the species has occured.}}$$

Frequency Classes : Frequency percentages can be arranged into the following five categories based on Raunkier's (1934) classification.

Frequency Class	Frequency Percentage
A	10-20
B	21-40
C	41-60
D	61-80
E	81-100

4. To Study the vegetation by line transect method and to determine the frequency of individual speicies :

Requirements : Measuring tape, graph sheet, herbarium sheet cellotape.

Procedure : Run the measuring tape to a suitable length (1 metre). and note down the species touching the tape. This length may be called one unit of a line transect. Run several such line transects randomly in different sites in the aea of study and note down the number of species . In addition count the number of individuals of each species touching the tape in each unit of the line transect. Tabulate the data as shown in the following table.

Name of species	Units line transect 1 2 3 4 etc	Total No. of transects in which the species present	Total No. of line transects studied	Frequency Percentage	Frequency Class
A	* + + + + (3) (2) (4) (1)				
B	- + + - (3) (2)				
C	- + - + (1) (2)				
D etc.					

* + indicates presence of species (not number of indivudals)
- indicates absence of species

** Value in parenthesis shows the number of individuals of particular species (A, B etc) in a given transect.

Calculate the frequency percentage for each species as in quadrat method.

Density of each species also may be calculated as in quadrat method.

5. To study the vegetation by point frame method and to determine the frequency of individual speicies.

Requirements : Point frame, graph sheet, herbarium sheet, cellotape etc.

Procedure : The point frame apparatus consists of a wooden frame (see figure) in the form of a rectangular bracket. The long horizontal arm has a number of holes to accomodate pins (nails).

Place the point frame apparatus (with 10 pins) in the field. The area occupied by the apparatus constitutes one sampling unit. Place the apparatus randomly in several places in the field and note down the number of species touching the pins. Also note down the number of individuals in each species touching the pins. Tabulate the data as in the case of transect method (substitute one unit point frame for one unit of line transect) and calculate the frequency percentage, frequency class and density.

6. To study the vegetation by Biological Spectrum Method

Requirements : Scalpel, scissors, graph sheet, spatula graph sheet, herbarium sheet, cellotape etc.

Procedure : In a given area of study examine each species as to the location of pernneating buds (or seeds) on the plant body. Note down the number of species. Now categorize the various species into the following life form types based on RaunKaier's classification as follows

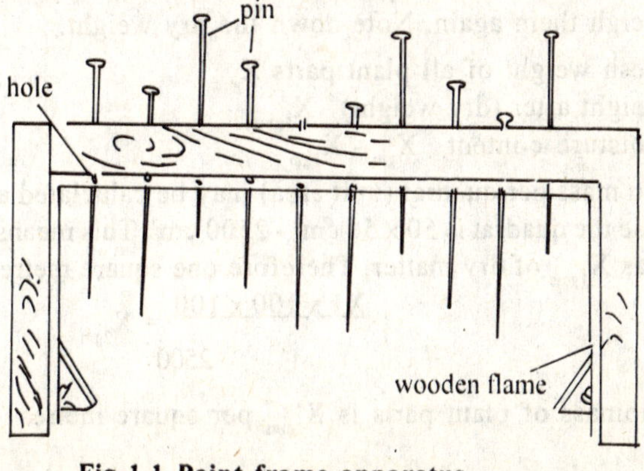

Fig 1.1 Point frame apparatus

1. Phanerophytes - Trees, shrubs with buds atleast 25 cms above ground.
2. Chamaephytes - Prostrate plants with buds on the surface or upto 25 cm above ground
3. Hemicryptophytes - Perennial plants with buds above or little below the soil surface.
4. Geophytes - Plants Rhizomatous or have under ground buds
5. Therophytes - Seasonal plants

Calculate the percentage value for each life form against the total life form in the area. With the help of a graph sheet plot the various life forms on the X axis and their percentage on the Y axis. Either a curve or histograms can be drawn on the graph sheet. This will provide the biological spectrum of the area.

Biological Spectrum may be defined as the distribution percentage of various life forms in a given area :

7. *To determine the biomass of green plants in a given area*

Requirements : Quadrat of known size, polythene bags, spatula, hot air oven etc.

Procedure : Lay down a number of quadrats randomly in a given area. Collect the plants in the quadrats as follows :

1. Cut all the aerial parts and put them in a polythene bag.
2. Collect all the dead, fallen leaves, twigs etc and put them in another bag.
3. Dig out all the underground parts of plants clean them of soil and collect in another bag.

Weigh each portion (1, 2, 3) separately and note down the fresh weight. Now put them separately in an oven at 90°C for 24 hours and weigh them again. Note down the dry weight.

a. Fresh weight of all plant parts X_{gms}
b. Weight after (dry weight) X_{1gms}
C. Moisture content $X_{gm} - X_{1gm}$

Bio mass per quadrat (unit area) may be caluclated as follows. Suppose the quadrat is 50 x 50 cm² - 2500 cm². This means 2500 cm² area has X_{1gms} of dry matter. Therefore one square metre will have

$$\frac{X1 \times 100 \times 100}{2500} = X_{2gm}$$

The biomass of plant parts is X_{2gms} per square metre

7. *To determine the vegetational cover in an area*

Requirements : Measuring tape, scissors, Vernier Calliperse, paper etc.

Procedure : Run the measuring tape for specific lengths at several places. Note down the length of plants touched by the tape.

Cut a transverse section of some plants at the ground surface. Measure the diametre of the cut end with the help of vernier scale. Tabulate the data as follows.

Name of Species	Units of tape 1 2 3 4 5 etc	Total
A B C etc.		
Total Number of Species	Length of tape for individual Plants	Total Length of all individuals

From the data above calculate the total length of tape (number of units x length of each unit). Covered by all the species and percentage of of total length of tape covered by each species. This gives the % vegetational cover of the area.

Name of species	Diametre No. of Individuals 1, 2, 3, 4 etc.	Total	Average
A B C etc			

Calculate the basal area as follows

basal area = πr^2 where

r = radius

π = 3.14

Average basal area for each species may be calculated by totalling the basal area of all species and dividing it by the number of species.

2

STUDY OF SOIL

Requirement : Polythene bag, pick axe, field lens, forceps, mesh of different sizes of pores.

Procedure : Collect soil samples from various sites in polythene bags. Examine the dry sample by a field lens. Examine a moist sample also. Squeeze the soil sample under thumb and feel the texture.

Categorize the soil as follows :

1.	Particles very large	Gravel
2.	Individual grains may be seen and felt between fingers. Do not adhere when	wet Sand
3.	Individual grains are seen and felt, form a cast when moist but fall apart when squeezed	Sandy loam
4.	Dry soil mellow with a gritty feel. Moist soil forms a cast which does not break while handling.	Loam
5.	Dry, appears cloudy, soft feeling, forms a cast (both dry and wet) of plastic material when wet	Silt loam

6. Hard when dry but forms cast which does not break while handling. Moist soil when pressed forms thin ribbon — Clay loam

7. Very hard lumps when dry, sticks to surface when moist; forms long ribbon when squeezed — Clay

By using meshes of different pore size also soil particles can be classified.

Diameter of the Particle in mm	Classification of particle
Less than 0.002	Clay
0.002-0.02	Silt
0.02-0.2	Fine sand
0.2-2.0	Coarse sand
2.0-5.0	Fine gravel
More than 5.0	Coarse gravel

2. *Study of soil temperature*

Requirements : Soil thermometer (ordinary thermometer also can be used) Iron rod etc.

Procedure : (i) Dig a hold of desired depth with the help of an iron rod and directly measure the temperature by inserting an ordinary thermometer. Record the reading after about 15 minutes.

(ii) Soil thermometers which have a steel end near the mercury bulb can be directly pushed into the soil to the desired depth and temperature measured.

3. *Measurement of soil pH*

Requreimennts : Soil samples, water, barium sulphate, Universal indicator or pH paper.

Procedure : Collect different samples of soil in separate test tubes. To each test tube add about 5 ml of barium sulphate and 15 ml of water. Keep the solution aside untill a clear supernatant fluid is formed. Take few drops of this on a porcelain tile and add few drops of universal indicator. Note the change in colour and compare it with the colour chart for different pH values pasted on the bottle. Alternatively dip a piece of pH paper in the fluid, note the change in colour and compare it with the colour chart.

4. *To determine the moisture content (%) in soil*

Requirements : Soil samples, hot air oven, china balance etc.

Procedure : Place soil samples of known weight in a hot air oven at a temperature of 105-110°C for 24 hours. Weight the soil sample before keeping them in hot air oven and also weigh after heating . Calculate the moisture percentage as follows.

1. Weight of soil before heating - X_{gms}
2. Weight of soil after heating for 24 hrs X_{1gms}
3. Moisture content X_1 - X
4. % moisture content = $\frac{X-X_1}{X_1}$ X 100

5. *Water holding capacity of soil*

Requirements : Soil samples, water, seive, oven, dessicator, balance etc.

Procedure : Pick a number of small holes in the base of the tin box. Fill the box with air dried and seived soil. Add water so that the soil gets saturated. Keep the tin in a slanting position and hang it to a stand with the help of string. Extra water comes out of the perforations at the base. Allow it to stand for a while. When water drops stop coming out remove the soil and weigh immediately. Note down the weight. Afterwards keep the soil in a hot air oven and dry it at 105°C for 48 hrs. Cool the soil sample in a dessicator and record the weight.

Observations and Conclusion : The difference in the weight in soil samples before and after heating gives the amount of water in the soil or water holding capacity of the soil. Calculate the water holding capacity per gram of soil. Also calculate the percentage of water retained by the soil. This is generally refered to as the field capacity of the soil.

6. *Calculation of carbonate content, nitrate content and base dificiency of soil samples.*

Requirements : Soil samples HCL, diphenylamine, H_2SO_4 ammonium thiocyanate, H_2O_2, water, china clay, tiles, test tubes etc.

Procedure :

(a) **Carbonate :** Take a little amount of soil sample in separate test tubes. Add a few drops of Hcl to each of the tubes. Observe the rate of effervescence. Compare the rates in different soil samples. Higher rate of effervscence indicates higher carbonate content.

(b) **Nitrate content :** Prepare a soil solutions of different samples by adding 5 parts water to one part water. Take a few drops of soil solution on a porcelain plate. Add a few drops of diphenylamine and a few drops Conc H_2So_4. Blue colour formation indicates the presence of nitrates. Higher intensity of colour indicates higher nitrate content. Different soil samples may be compared based on the intensity of colour.

(c) **Base defeciency :** Take a pinch of soil sample in a tube, add saturated alcholic solution of ammonium thiocynate and shake vigourously. Allow the contents to settle. Coolect the supernatant and to it add a few drops of H_2O_2 and observe the intensity of red colouration that develops. Compare the degree of colouration in different soil samples. Higher intensity of colour indicates increased dificiency of replacable bases.

7. Test for the presence of Inorganic Salts

Requirements : Soil samples, distilled water, barium chloride solution, Hcl, H_2So_4, Nitric acid ammonium nitrate, ammonium molybdate, silver nitrate, beakers, test tubes, conical flask etc.

Procedure : Take 200 gms of soil sample in a conical flask and add 500 ml of water. Shake vigourously and keep it aside overnight to allow soluble salts to dissolve; filter the water and collect the filterate (aqueous soil extract) conduct the following tests.

(a) **Chloride :** Take 10 ml of aqueous soil extract in a test tube and add 10 ml of 0.1 N H_2So_4 to neutralize the carbonate and bicarbonate content. Now add silver nitrate to the mixture. Development of white precipitate indicates the presence of chloride.

(b) **Phosphate :** Take 10 ml of aqueous soil extract. Add a few drops of ammonium molybdate solution, Conc. nitric acid and ammonium nitrate. Development of yellow colour indicates the presence of phosphates.

(c) **Sulphate :** Take 10 ml of aqueous soil extract in a test tube and add 5 ml of Conc. HCl and boil cool and add a few ml of $BaCl_2$ solution. Development of white precipitate indicates the presence of sulphate.

3

STUDY OF ATMOSPHERE

1. *Measurement of temperature, relative humidity and light intensity of air (atmoshphere).*

Requirement : Maximum - minimum therometer, and wet thermometer (psychrometer) photometer etc.

Procedure : Temperature of the atmosphere can be measured by means of 'U' shaped thermometer in which one arm records the minimum and the other maximum temperature. The 'U' Shaped thermometer may be kept in different places like open air, shade etc and the temperature may be recorded.

Relative Humidity : This is recorded with the help of psychlometer also know as dry and wet thermometer (see fig). THe psychlometer has two thermometers - one dry and the other wet. The bulb of wet thermometer is always kept moist by dipping it in moist cotton. Evaporation of water due to normal temperature of the atmosphere causes cooling of the bulb of wet thermometer and lowers the temperature. Dry thermometer records the normal temperature. Note the difference in temperature between dry and wet thermometers. Relatiive humidity may be calculated by using a conversion table. If the relative humidity is 100% there will be no difference in temperature between the dry and wet thermometer.

2. *Measurement of light intensity*

Requirements : Photometer (Lux meter)

Procedure : Photometer consists of a photo electric cell and an amperemeter (microammeter). The instrument has a filter to guard against intense light. Expose the apparatus and read the value with the deflection of the needle. The photo electric cell on exposure to light converts it into electricity and the current generated is measured in the ammeter in units called lux units.

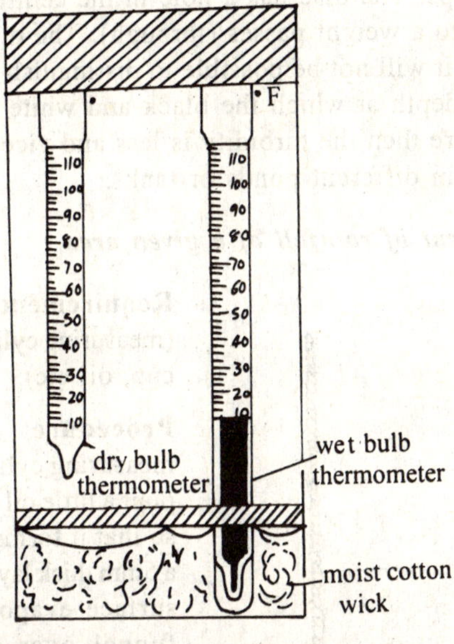

Fig 3.1Dry and Wet Thermometer

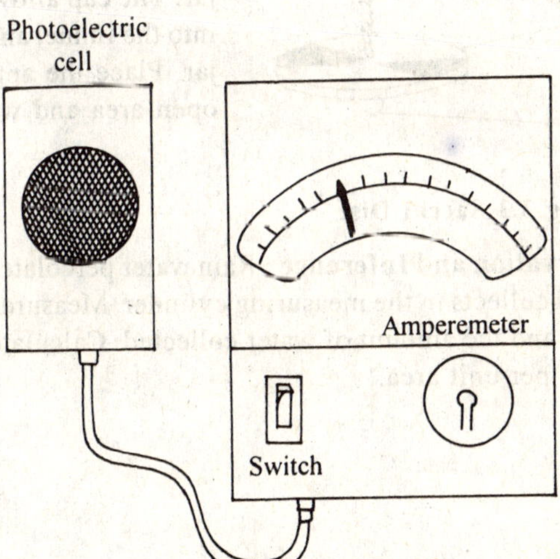

Fig 3.2 Light Meter

3. *Measurement of turbidity of water :*

Requirements : Secchi disc

Procedure : Secchi disc is a circular piece of metal having black and white strips. The disc has a hole in the centre through which a thread tied to a weight passes (through). The disc is lowered to a level where it will not be possible to distinguish black and white strips. If the depth at which the black and white strips lose their identity is more then the turbidity is less and vice versa. This can be compared in different ponds or tanks.

4. *Measurement of rainfall of a given area :*

Requirement : Rain guage (measuring cylinder, funnel, jar, cup, oil etc).

Procedure : Keep the measuring cylinder in a jar and pour a little oil into the cylinder so that it forms approximately a 5mm thick layer. This prevents surface evaporation. Place a funnel over the measuring cylinder and fix a cap over the jar. The cap allows water only into the funnel and not into the jar. Place the apparatus in an open area and watch.

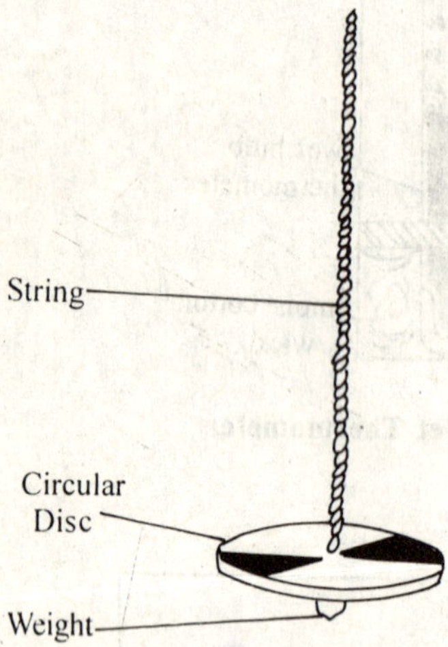

Fig 3.3 Sacchi Disc

Observation and Inference : Rain water percolates through the funnel and collects in the measuring cylinder. Measure the diameter of the cap and the amount of water collected. Calculate the amount of rainfall per unit area.

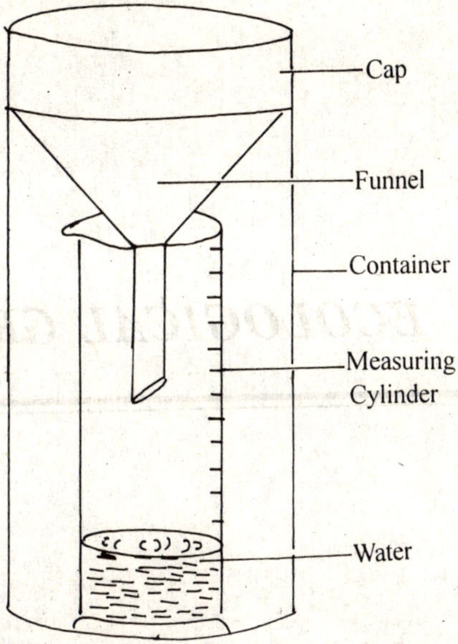

Fig 3.4 Rain gauge

4

ECOLOGICAL GROUPS

Groups of plants belonging to one or more species that grow together in a particular area constitute an ecological community or a group.

For instance trees shrubs, climbers etc that grow in a forest form a forest community. Similarly plants that grow in a pond form a pond community. Likewise, plants that grow in desert form a desert community. Based on the environment they inhabit warming (1985) has identified eight ecological groups or communities.

HYDROPHYTES

Plants that grow in water or in soil covered with water belong to this group. Hydrophytes have a number of morphological and anatomical adaptations. These are

Morphological Adaptations :

1. **Dimorphic leaves :** Some of the plants which are partially submerged have dimorphic leaves. The underwater leaves are thin and highly dissected whereas the leaves above water are broad. This condition is known as heterophyllly. Ex. *Limnophylla heterophylla, Trapa, Sagittaria sagittifolia* etc

2. **Formation of coiled pedicles and petioles :** In plants like *Vallisnaria* the pedicles and petioles are coiled so that the flowers and leaves will be always on the surface of water but do not ofter resistannce to water currents.

3. **Development of water Proofing device :** To prevent decay a waxy coating or mucilaze covering is developed on the surface to repel water.

4. **Roots :** Highly reduced, divided into thin fibrels or even absent (ultricularia). Root hairs and root caps are also absent very often.

5. **Stem :** Very thin, delicate and transparent.

6. **Guttation :** Many of the plants remove water by way of guttation.

Anatomical Adaptations

1. **Epidermis :** Cells are thin walled, contain chloroplasts

2. **Aerenchyma :** Parenchyma tissue has a large number of air cavities to store air which helps both in respiration and buoyancy (floating).

3. **Cuticle :** Absent or even when present very thin

4. **Mechanical Tissues :** These are very poorly developed. Sclerenchyma and xylem which are the principal lignified tissues are very little in extent.

5. **Stomata :** Absent in submerged leaves, but present in the upper epidermis of floating leaves

6. **Secondary growth :** Takes place very rarely.

7. **Chlorenchyma :** Very well developed.

Classification of Hydrophytes

1. **Floating hydrophytes :** Plants float on the surface of water. There are two kinds of surface hydrophytes - free floating and fixed. In the former the roots are suspended freely in water (e.g. *Eichhornia, Pistia, Salivnia* etc) Fixed floating hydrophytes have their roots fixed to the underwater soil. They occur in shallow water (e.g *Marsilea, Nymphaea* etc).

2. **Submerged hydrophytes :** They are completely under water; may be suspended freely in water (e.g. *Elodea*) or may be rooted to soil underneath the water (e.g *Vallisnaria*).

3. **Amphibious Hydrophytes :** They grow in shallow or muddy waters. Roots and parts of their stems are embedded in water while their branches and leaves are often aerial (e.g. Typha, Alisma etc). We will now study in detail morphological and anatomical adaptations in a few exapmles.

Fig 4.1Hydrilla - Habit

Hydrilla verticellata (Hydrocharitaceas)

Morphology

1. It is a commonly occuring water plant in ponds.
2. It is a submerged hydrophyte, but rooted to the under water soil.
3. Root system is adventitious.
4. Stem is transparent, flexible and soft.
5. Leaves are small, thin and transparent. Phyllotaxy is whorsed.

6. Mucilage covers all parts of the plant body to prevent excessive excessive absorption.

Transverse Section of stem

1. Epidermis is single layered made up of isodiametric thin walled cells.
2. Cuticle if any is very thin
3. These is a single layer of parenchymatous hypodermis
4. Cortex is made up of aerenchyma. Air chambers are irregular in size and shape and are separated. by one layered partititon of cells. Cortical cells are spherical or oval and thin walled. Air cavities are larger in size towards periphery but smaller towards the centre.
5. Internal to the aerenchymatous cortex is the endodermis surrounding the stele.
6. The stele is highly reduced. Xylem is absent and is replaced by an air cavity. Surrounding the air cavity is phloem.

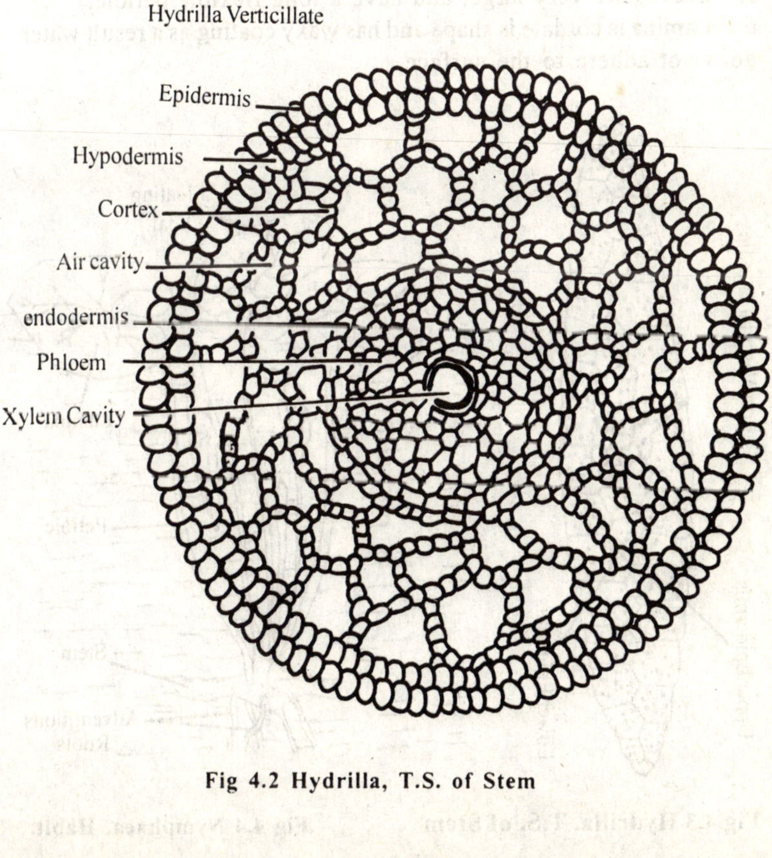

Fig 4.2 Hydrilla, T.S. of Stem

Transverse Section of Leaf

1. Leaf is dorsiventrally flattened and reduced in size.
2. The upper and lower epidermal layers are made up of a single layer of cells each. There is no cuticle but there is deposition of mucilage.
3. There are no stomata
4. Vasculature is highly reduced. There is no bundle sheath
5. Mesophyll is also highly reduced in quantity and undifferntiated
6. All cells including epidermis are chlorophyllous

Nymphaea (Nymphaeaceae)

Morphology

1. It occurs in shallow waters
2. Plant is a submerged rooted hydrophyte
3. Stem is rhizomatous growing horizontally underneath the soil.
4. Roots are adventitious.
5. Leaves are very large, and have a long flexible petiole.
6. Lamina is cordate is shape and has waxy coating as a result water does not adhere to the surface.

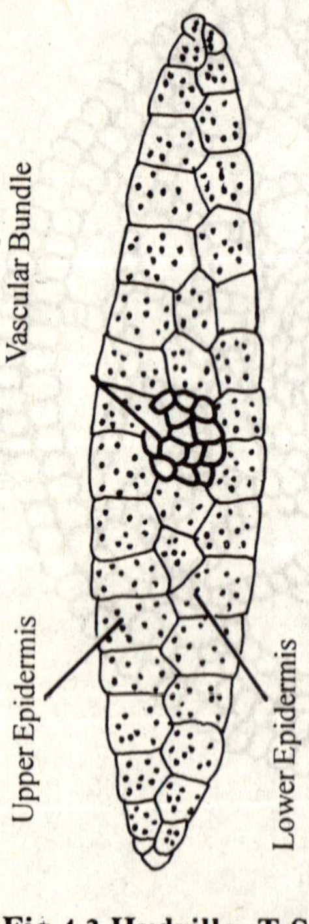

Fig 4.3 Hydrilla, T.S. of Stem

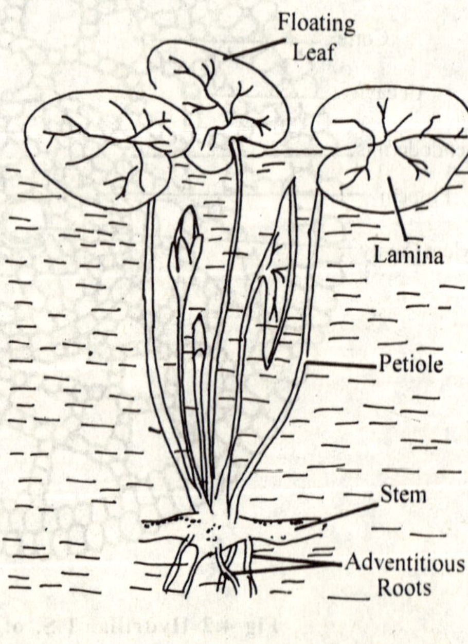

Fig 4.4 Nymphaea, Habit.

Transverse Section of Petiole

1. Outermost layer is epidermis. It is made up of thin walled barrel shaped cells. Cells are compactly arranged. There are a number of multicellular hairs arising from the epidermis. A thick coating of mucilage is found on the epidermal walls.

2. Internal to the epidermis is a few layers of collenchyma constituting the chief mechanical tissue. (Collenchyma is a living mechanical tissue whose corner walls are thickened). The collonchyma constitutes the hypodermis.

3. Next to the hypodermis is the ground tissue. Immediately next to hypodermis there are a few layers of compactly arranged cells (parenchyma) with small intercellular cavities. Next to this is the aerenchyma with large air spaces separated by uniseriate partitions.

4. Branched or stellate sclerids are found protuding into the air cavities.

5. Vascular bundles are reduced. They are scattered in the ground tissue being present mostly in the outer zone.

6. In the vascular bundle xylem is represented by a lacuna. Phloem elements are present in their normal condition.

Transverse Section of leaf (lamina)

1. The lamina is dorsiventrally flattered (bifacial)

2. Upper epidermis is single layered. waxy cuticle is deposited above the epidermis.

3. Stomata are found only in the upper epidermis (epistomatic).

4. Lower epidermis is also made up of a single layer of cells. There is no cuticle on lower epidermis.

5. Mesophyll is differntiated into upper palisade parenchyma and lower spongy parenchyma.

6. Palisade parenchyma is 2 layered. Cells are compact. This is the chief phtosynthetic tissue in the leaf.

7. Spongy parenchyma cells have very large air cavities between them.

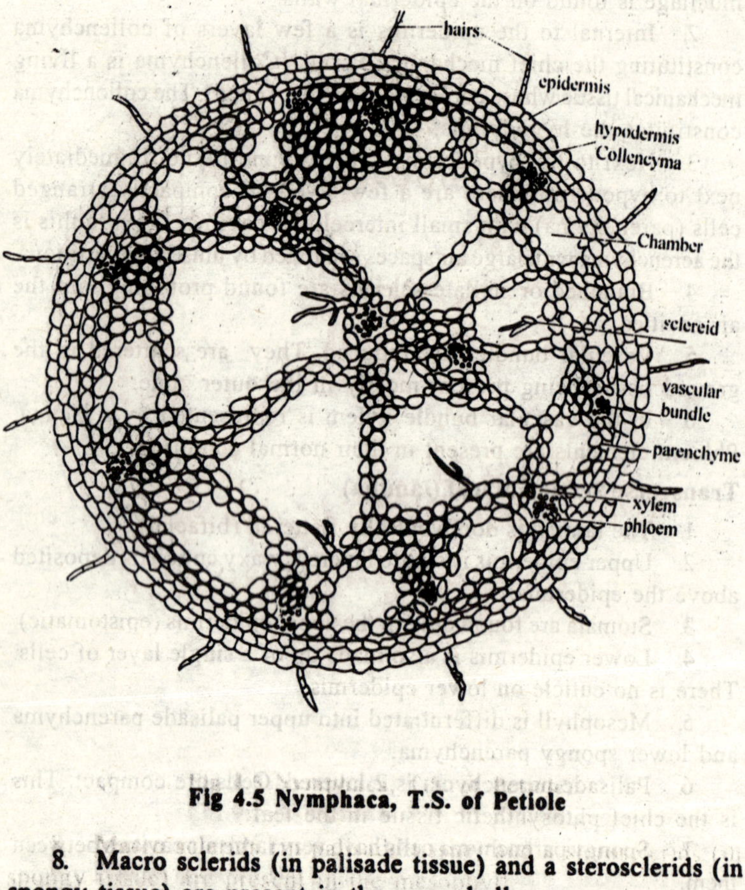

Fig 4.5 Nymphaca, T.S. of Petiole

8. Macro sclerids (in palisade tissue) and a sterosclerids (in spongy tissue) are present in the mesophyll.

9. Vascular bundles are poorly developed. They contain protoxylem cavities. Phloem is well developed.

Trapa (Trapaceae)

1. Trapa is a free floating hydrophyte grows in open tanks and lakes.

2. Leaves are dimorphic floating and submerged. Floating leaves are large. They have a swollen petiole to help in buoyancy.

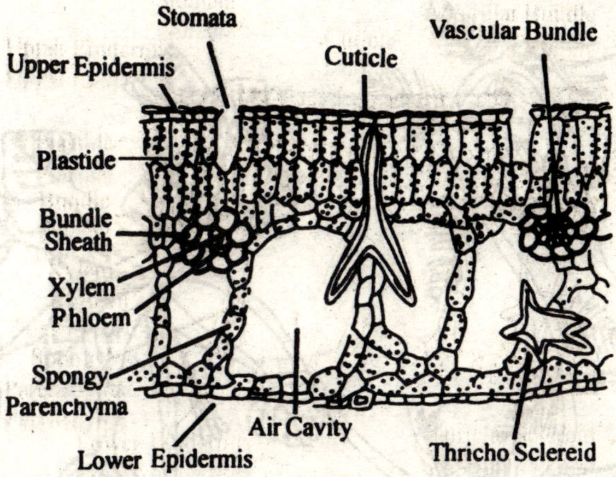

Fig 4.6 Nymphaea, T.S. of Lamina

3. Submerged leaves arise in pairs at each node of the stem. They are highly dissected.

4. Stem is also submerged delicate and somewhat spongy.

5. Roots are unbranched and arise from the nodes.

6. Flowers are aerial but after pollination the pedicel bends down into water and the development of the fruit is under water.

Transverse Section of Stem

1. There are three regions in a transverse section epidermis, cortex and stele.

2. Epidermis is single layered. Multi cellular hairs arise from the epidermal cells.

3. Below the epidermis is a few layers sclerenchyma constituting the hypodermis.

Fig 4.7 Trapa, Habit.

4. Cortex is made up of parenchyma with large air cavities.
5. sphaerarophides are found in the cortical cells
6. Internal to the cortex is the endodermis.
7. Pericycle is found internal to the endodermis
8. Vascular bundles are many in number arranged in the form of a ring surrounding a central pith.
9. Xylem is poorly developed. Phloem is found both external as well as internal to xylem.
10. Pith is wide and consists of parenchyma cells.

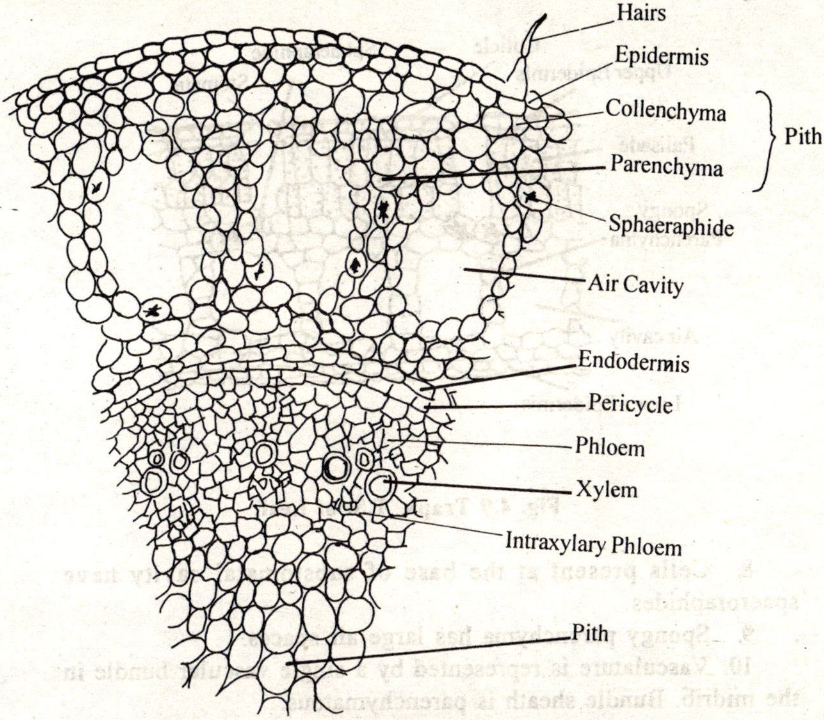

Fig 4.8 Trapa, T.S. of Stem (A part enlarged)

Transverse Section of Leaf

1. Leaf is dorsiventrally flattened.

2. Upper epidermis consists of isodiametric cells. Cuticle is deposited above the epidermis.

3. Stomata are found only in the upper epidermis (leaves are epistomatic).

4. Lower epidermis is single layered and the cells are uninterrupted.

5. Mesophyll is distinguished into palisade and spongy parenchyma

6. Palisade cells may be in one or two layers below the upper epidermis. They have a number of chloroplasts.

7. Stomata in the upper epidermis lead to substomatal cavities present between the palisade cells.

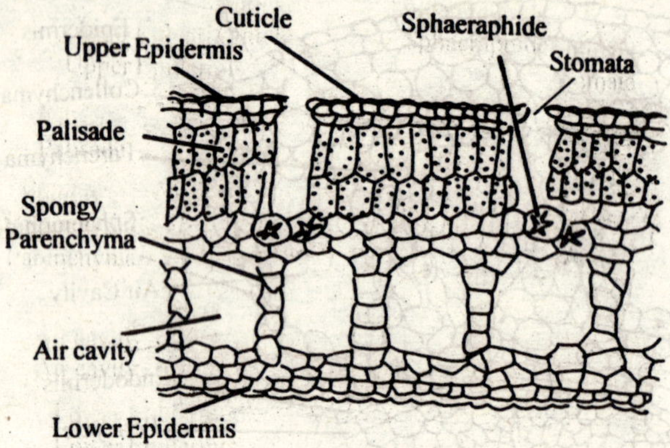

Fig. 4.9 Trapa, T.S. of Leaf.

8. Cells present at the base of substomatal cavity have spaeroraphides.
9. Spongy parenchyma has large air spaces.
10. Vasculature is represented by a single vascular bundle in the midrib. Bundle sheath is parenchymatous.
11. Xylem is reduced to cavities. Phloem is well developed.

Potomogeton (Hyd:acharitaceae)

Morphology

1. The plant is a submerged rooted hydrophyte growing is shallow ponds and lakes.
2. Flowering branches are aerial.
3. Shoot is horizontal with vertical branches. Branches are thin delicate and long.
4. Roots are adventitious arising at the nodes
5. Leaves are semitransparent thin, thread like with a distichous phyllotaxy (arranged in two rows)
6. Mucilage covers all the underwater parts of the plant

Transverse Section of Stem

1. Transverse Section shows three regions - epidermis, cortex and stele.

2. Epidermis is single layered made up of barrel shaped cells. Cuticle is deposited above the epidermis.

3. Internal to the epidermis is a wide aerecnymatous cortex consisting of large air cavities separated by one celled partitions. Small patches of sclerenchyma are found in the cortex.

4. Endodermis is well developed while pericycle is indistinct

5. Xylem is reduced to a cavity while phloem is well developed.

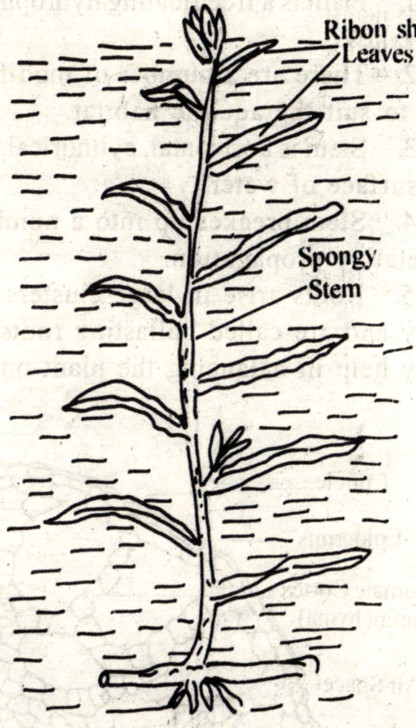

Fig 4.10 Potomogeton, Habit

Transverse Section of Leaf

1. Leaf is dorsiventrally flattened
2. Upper and lower epidermis are single layered arranged compactly.
3. Stomata are found only in the uppder epidermis
4. Each stoma leads to a sub stomatal cavity
5. Mesophyll is differentiated.
6. Palisade parenchyma is loosely arranged, so also spongy parenchyma.
7. Vascular bundle is surrounded by bundle sheath.
8. Xylem is highly reduced. Phloem is somewhat well developed.
9. At the margin leaf is just 3 celled in thickness.

Eichhorinia (Potenderiaceae)

Morphology

1. Plant is a free floating hydrophyte growing profusely in ponds and lakes.
2. There are a number of modifications in the root, stem and leaf to suit the aquatic habitat.
3. Stem is horizontal, cylindrical, spongy and is found just above the surface of water.
4. Stem breakes up into a number offsets which bring about vegetative propagation.
5. Roots arise in large clusters at the nodes, black in colour hairy and are called ballasting roots.

They help in balancing the plant on water.

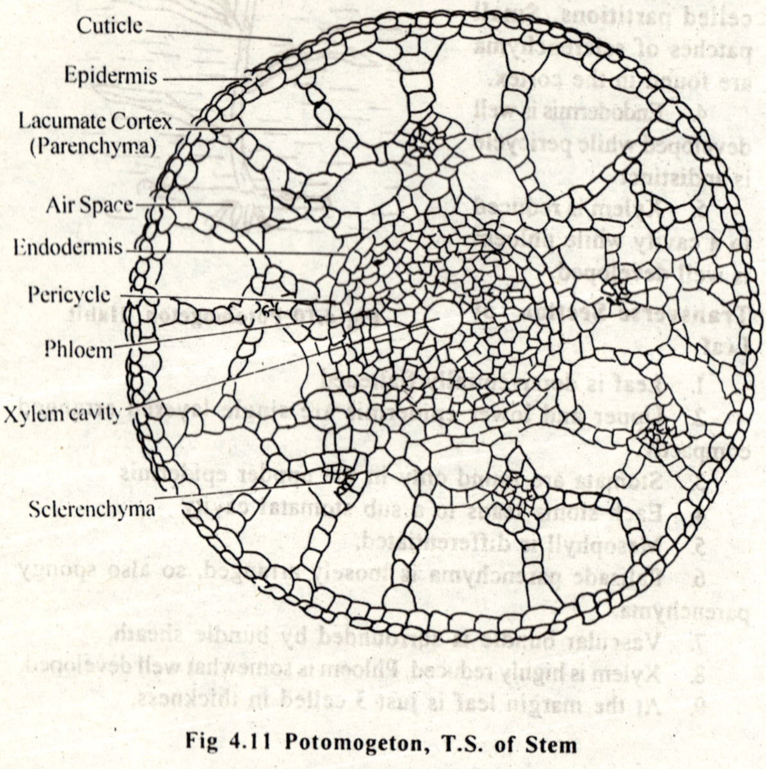

Fig 4.11 Potomogeton, T.S. of Stem

6. Root caps are basent, instead large root packets are found.
7. Leaf has a spongy and bulbous petiole called phyllode. It helps in buoyancy.

8. Lamina is shiny due to the deposition of waxy material.
9. Pinkish flowers arise in clusters at the node

Transverse Section of Petiole

1. A T.S. shows epidermis and ground tissue in which the vascular bundles are scattered.

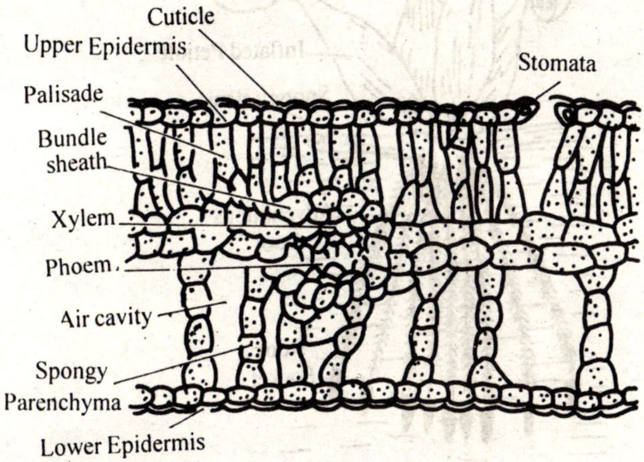

Fig 4.12 Potomogeton, T.S. of Leaf

2. Epidermis is made up of a single layer of thin walled cells. Cuticle is deposited over the epidermis.
3. Internal to the epidermis is the hypodermis made up of a few layers of compactly arranged cells. Cells are chlorophyllous.
4. Internal to the hypodermis is the ground tissue made up of aerenchyma with large air chambers separated by one cell thick partitions.
5. Dispersed in the ground tissue are found a number of vascular bundles. Xylem is poorly developed while phloem is well developed.

Transverse Section leaf (lamina)

1. Leaves are dorsiventally flattened
2. Both upper and lower epidermis are made up of a single layer of cells each.
3. Cuticle is deposited above the lower epidermis
4. Stomata are found in both the upper and lower epidermis (amphistomatic)

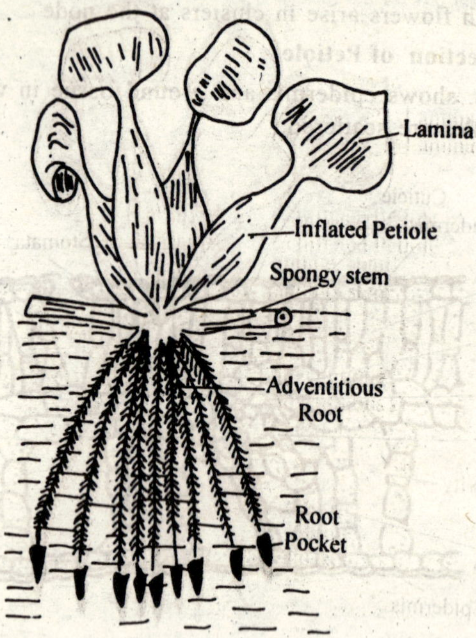

Fig 4.13 Eichhornia, Habit

5. Mesophyll is differentiated.
6. Palisade parenchyma is one or two layered
7. Spongy parenchyma has large number of air cavities.
8. Vascular bundles have a prominent bundle sheath. Xylem however is poorly developed while phloem is well developed.

Transverse Section of Root

1. A Transverse Section shows the three regions - epidermis, cortex and stele.
2. Epidermis is single layered made up of compactly arranged barrel shaped cells.
3. Internal to the epidermis is a few layered hypodermis followed by a broad aerenchymatous cortex.
4. Stele is surrounded by endodermis and pericycle
5. Vascular bundles are radial with exarch xylem. Xylem however is not well developed.
6. In the centre of the stele is present a parenchymatous pith.

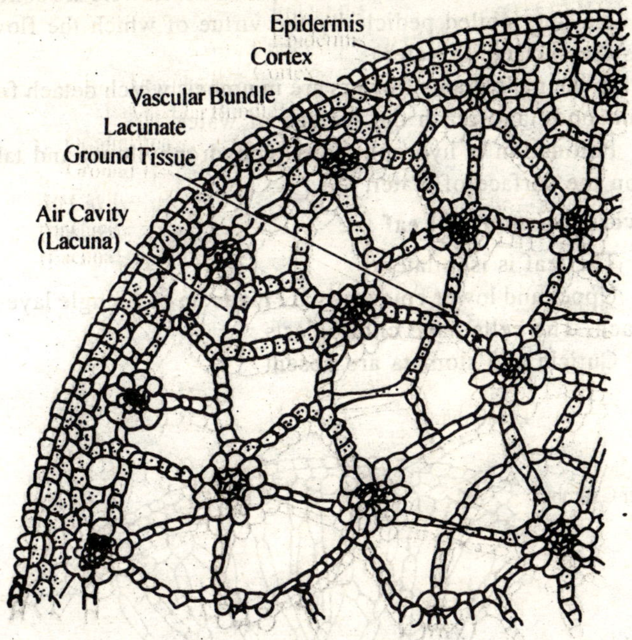

Fig 4.14 Eichhornia, T.S. of Petiole (A part enlarged)

Vallisnaria (Hydrocharitaceal)

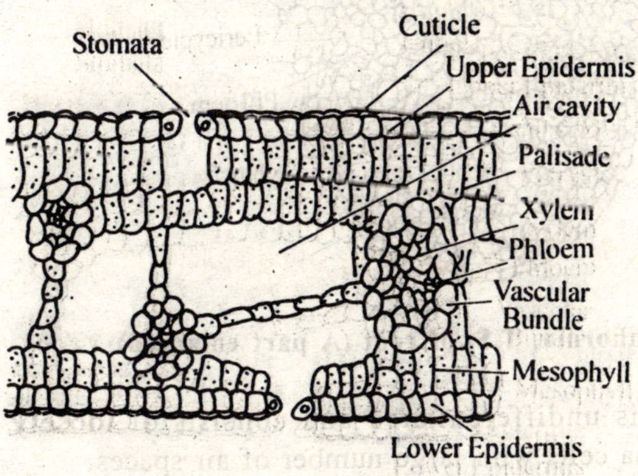

Fig 4.15 Eichhornia, T.S. of Petiole (A part enlarged)

1. Vallisnaria is a submerged rooted hydrophyte found growing in stationary or running water.

2. Roots are adventitious arising at the nodal regions of the horizontal under-ground stem.

3. Leaves are flat ribbon shaped and thin arising in a rosette fashion from the stem.

4. Plants are dioecious. In female plants the flowers are solitary and have a long coiled pedicle by the virtue of which the flower reaches the surface of water.

5. In the male plants flowers are in umbels which detach from the plant body and reach the surface of water.

6. Pollination is hydrophyllous (ephydrophyllous) and takes place on the surface of water.

Transverse Section of Leaf

1. The leaf is isobilateral
2. Upper and lower epidermis are made up of a single layer of cells each. The cells have chloroplasts.
3. Cuticle and stomata are absent

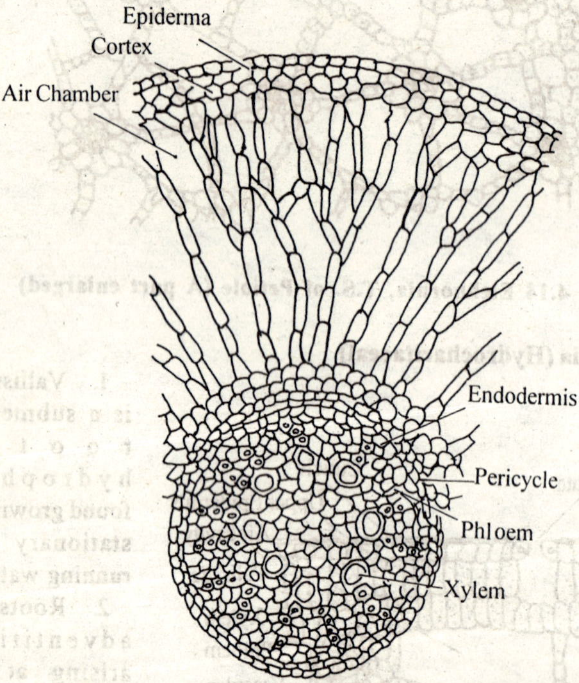

Fig 4.16 Eichhornia, T.S. of root (A part enlarged)

4. Mesophyll is undifferentiated and consists of loosely arranged parenchyma cells with large number of air spaces.
5. Vascular bundles are reduced, Xylem is represented by a cavity while phloem is comparatively well developed.

Jussieua (Onagraceel)

Morphology

1. Plant is a free floating hydrophyte found in lake, ponds and ditches

2. Stem is horizontal growing just beneath the surface of water. It is distinguished into nodes and internodes. Stem is soft and spongy.

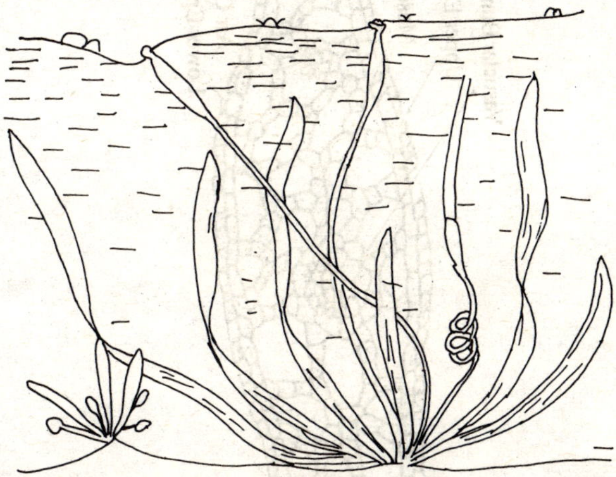

Fig 4.17 Vallisnaria, Habit

3. There are two types of roots - floating roots and ordinary roots. Ordinary roots are thin, fibrous and grow into water. Floating roots are white in colour spongy, soft short and stumpy arising in clusters of 3-4 at every node. They help in buoyancy.

4. In the anchorage roots, root hairs are absent, instead root packets are present.

Transverse Section of Stem

1. Transverse Section shows epidermis, cortex and stele.

2. Epidermis is made up of thin walled compactly arranged, radially elongated cells. Cuticle is absent

3. Next to the epidermis is the cortex divided into three zones. - outer cortex, middle cortex and inner cortex.

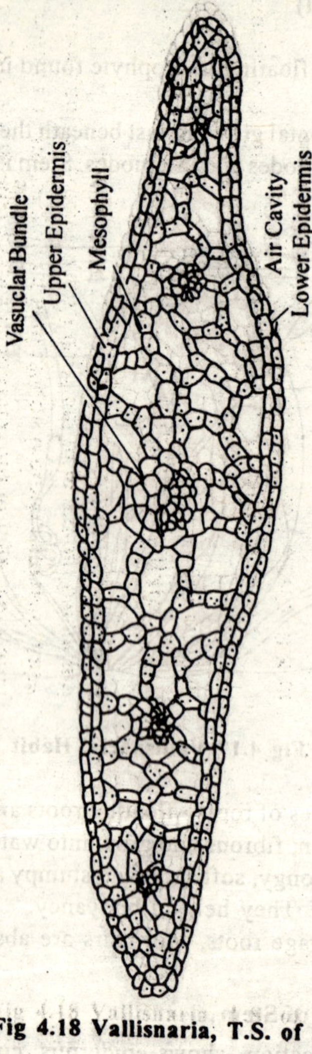

Fig 4.18 Vallisnaria, T.S. of Leaf

4. Outer cortex is composed of thin walled parenchyma cells with few intercellular spaces.

5. Middle cortex is aerenchymatous with large number of air spaces. The air spaces are separated by 3 - 4 celled wide partitions.

6. Inner cortex is made up of 2-3 layers of compactly arranged parenchyma cells.

7. Stele is surrounded by an indistinct endodermis.

8. Vascular bundles are many in number arranged in the form of a ring surrounding a central pith.

9. Vascular bundles are collateral and conjoint with poorly developed xylem but comparatively well developed phloem.

Ceratophyllum

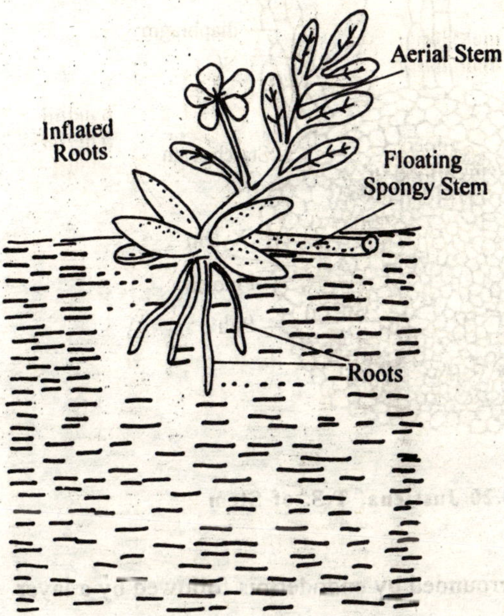

Fig 4.19 Jussieua, Habit

Morphology

1. Plants are submerged but rootless hydrophytes. They are feely suspended in water.

2. Stem is long, slender and delicate. Branches arise at the nodes.

3. Leaves are dichotomously dissected and arise in whorls at the nodal region. Margins of the leaves are serrate.

4. All the plant parts are covered with mucilage.

5. Mechanical breakage of the stem results in vegetative propagation.

Transverse Section of Stem

1. Transverse Section shows outer epidermis, middle cortex and the central stele.

2. Epidermis is made up of a single layer of thin walled compactly arranged cells. Cells are chlorophyllous.

3. Cortex is wide, chlorophyllous aerenchymatous with large number of air spaces.

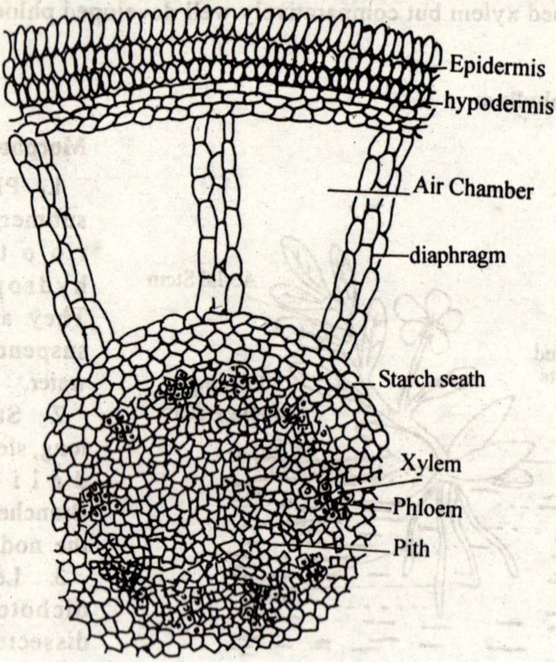

Fig 4.20 Jussiena, T.S. of Stem

4. Stele is surrounded by endodermis followed by a layer of pericycle.

5. Xylem is represented by an air cavity. Phloem is comparatively well developed.

Transverse Section of Leaf

1. The outline of cross section is almost circule.
2. Epidermis is single layered made up of compactly arranged thin walled chlorophyllous cells. Stomata are absent.
3. Mesophyll is undifferentiated composed of compactly arranged parenchyma cells. However 2-3 large air chammbers are present in the mesophyll.
4. Vasculare bundle is hihgly reduced and has poorly differentiated conducting tissue.

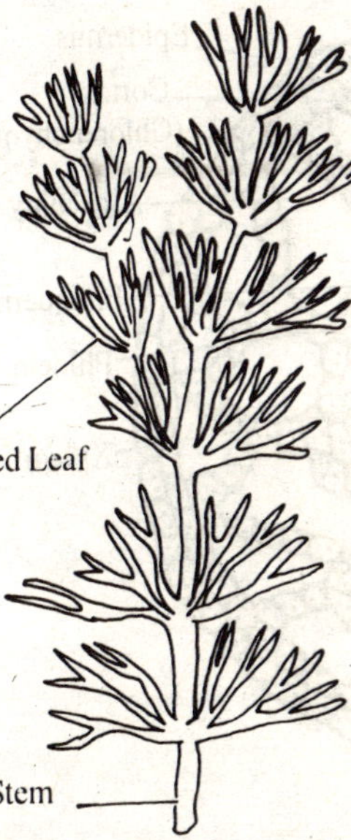

Fig 4.21 Ceratophyllum, Habit

Nelumbo (Lotus)

1. Plants are rooted hydrophytes with floating leaves.
2. Leaves have a wide lamina with a long cylindrical soft air filled petiole.
3. Petiole and surface of lamina is coated with wax to prevent excessive absorption of water.
4. Some leaves may also rise above the surface of water.
5. Stomata are present on the upper epidermis.

Pistia (Araceae)

1. Free floating hydrophytes found on water surface in ponds, lakes etc.
2. Stem is horizontal, cylindrical and greenish in colour
3. Branches of stem called offsets help in vegetative propagation.
4. At each node on the stem a rosette of leaves arise in a spiral
5. Roots are adventitious arising in groups from the lower surface of the stem at the nodal regions.
6. Roots do not have root caps, instead they have root packets.

Xerophytes

Plants that grow in places where the availability of water is less are called xerophytes. Some xerophytes are also called desert plants. Xerophytes develop several morphological and anatomical adaptations to economise water.

XEROPHYTES Morphological adaptations

1. Root system is well developed. It is both intensive and extensive for efficient water absorption.
2. Roots may be shallow but spread extensively when they are called surface feeders.
3. Roots are said to be deep feeders when they are not spreading but grow deeply into the soil.
4. Stems are stunted, thick and extensively branched.

5. In some cases stem gets modified into thick fleshy green structures called phylloclades or cladodes.

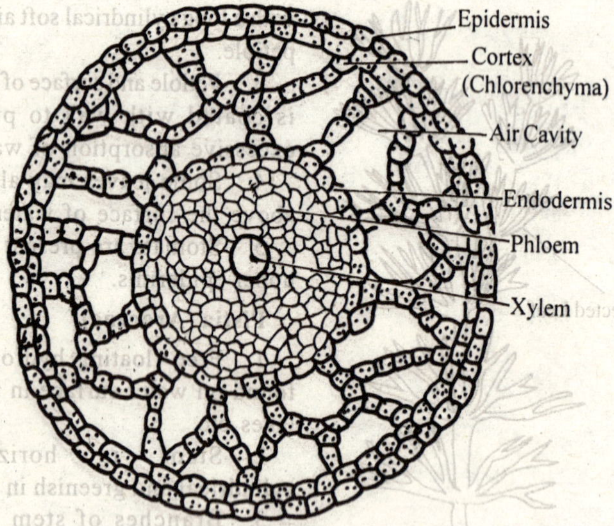

Fig 4.22 Ceratophyllum T.S. of leaf

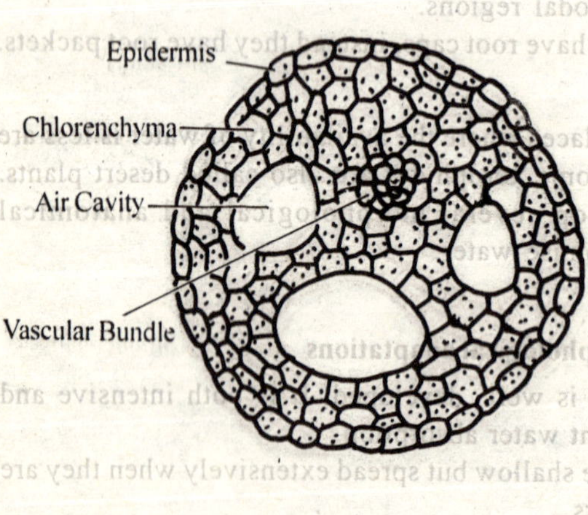

6. Stems may be underground in some cases or become hard and woody with thick barks.

7. Leaves are generally reduced, modified or totally eliminated.

Fig 4.23 Ceratophyllum T.S. of Leaf

8. Leaves may be modified into scales or spines which helps in checking transpiration as well as protects the plants from grazing animals.

9. In some plants leaves remain in a vertical posture exposing very little surface to direct sun light.

10. Some of the plants have decidous leaves which fall off during summer.

11. Leaves when present have a waxy or shiny surface to reflect right.

12. Leaves may be leathery or may have hairy outgrowths on the surface.

13. Under xeric conditions leaves may roll up and open only on the availability of water. (e.g. *Selaginella rupestris*).

14. Xerophytes produce large number of perenniating buds on the stem to help in vegetative propagation.

Anatomical Adaptations

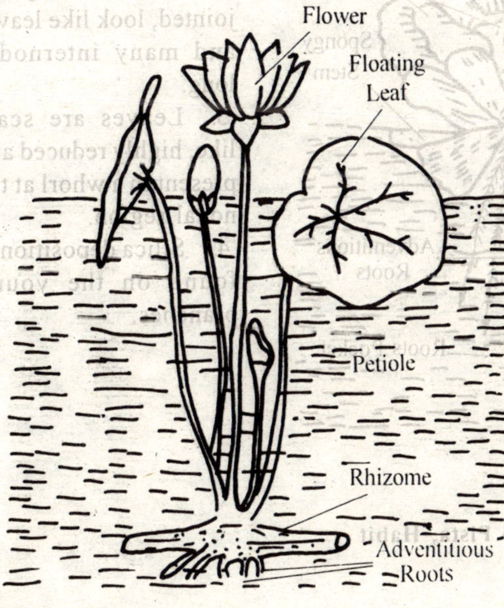

Fig 4.24 Nelumbo, Habit

1. Epidermis is well developed. Sometimes multiple epidermal layers are found.

2. Cuticle is very well developed.

3. Waxy or silica coating is often found on the aerial parts.

4. Hypodermal sclerencyma is often found beneath the epidermis in the leaves.

5. Stomata are reduced in number and usually leaves are hypostomatic

6. In some plants stomata are scotoactive (i.e. open only during night).

7. Stomata are sunken, often covered by hairy outgrowths.

8. Many stomata are often found together in a sunken cavity (as in *Nerium* leaf).

9. Palisade parenchyma is very well developed; often it is many layered.

10. Air spaces are less in spongy parenchyma.

11. In succulent stems water is mixed with mucilage to prevent evaporation.

12. Mechanical tissues such as sclerenchyma, xylem etc are very well developed.

13. In some grasses motor cells are bulliform cells are present to help in the rolling up of the leaves.

We shall now study few examples of xerophytes

Casuarina equisetifolia

Morphology

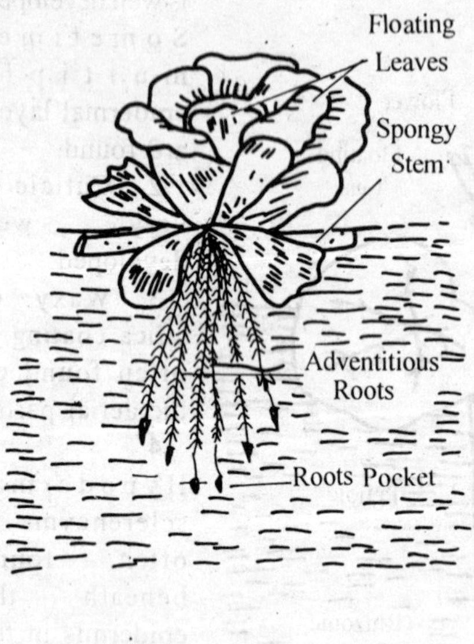

Fig 4.25 Pista, Habit

1. The plant is a tall tree.
2. Branches are green, jointed, look like leaves and many internodes long.
3. Leaves are scale like, highly reduced and present in a whorl at the nodal region.
4. Silica deposition is found on the young branches.

Transverse Section of Stem

1. Outline of the transverse section will be vaxy because of ridges and furrows.

2. Outermost layer is epidermis made up of a single layer of cells-cuticle is deposited above the epidermis.

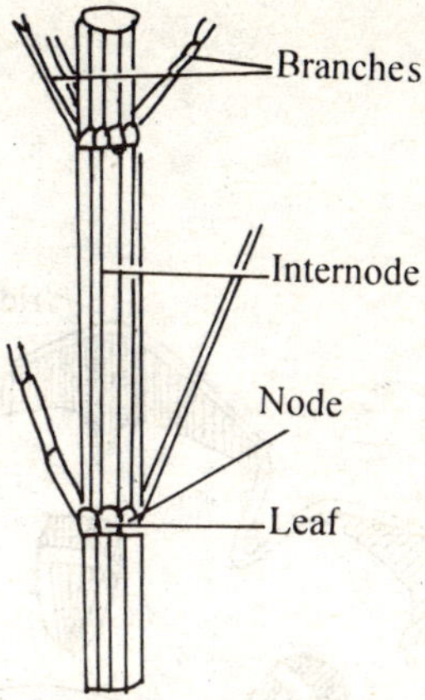

Fig 4.26 Casuarina, a branch

3. Stomata are sunken, restricted in distribution to the furrows.

4. Epidermal cells in the furrow have long multicellular hairs filling the cavity.

5. Next to the epidermis are found patches of sclerenchyma beneath the ridges.

6. Internal to the hypodermis the cortex is divided into ourter chlorenchyma and inner parenchyma

7. Chlorenchyma cells are arranged in many palisade layers constituting the chief photosynthetic tissue.

8. Embedded in the parenchymatous cortex are found a number of vascular bundles (one each below the ridge). These are called cortical vascular bundles and are collateral, conjoint, open with endarch xylem.

9. Next to the cortex is found the endodermis. Pericyle is represented by patches of sclerenchyma found between the endodermis and the stelar vascular bundles.

10. Vascular bundles are many arranged in the form of a ring surrounding a central pith. They are similar to the cortical vascular bundles in strcuture. Many phloem fibres are present in the vascular bundles.

Calotropis gigantea

Morphology

1. The plant body is a large herb or shrub.

2. All aerial parts are covered with white waxy material which acts as insulation against heat

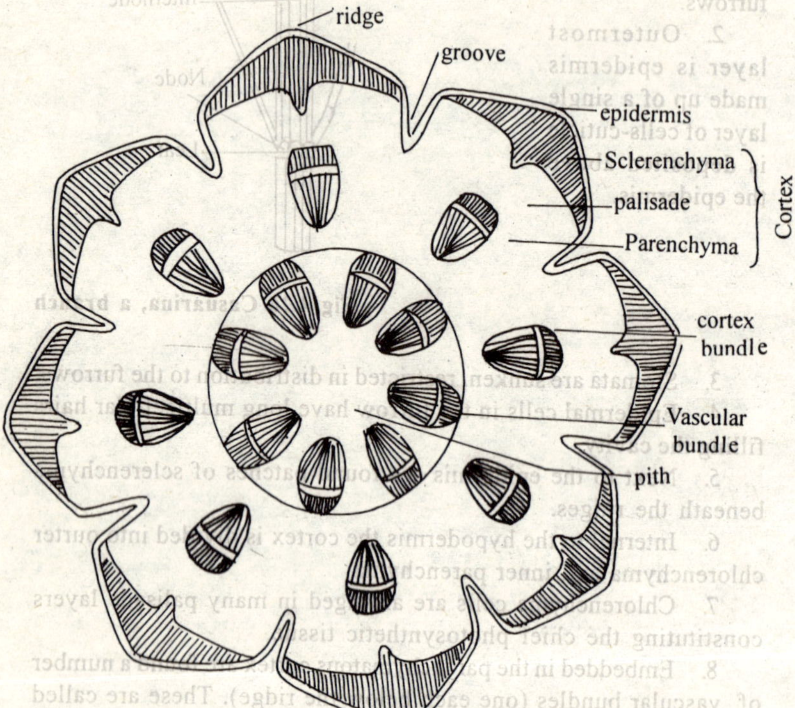

Fig 4.27 Casuarina, T.S. of Stem (Ground Plan)

3. Thick milky latex is present in the plant.

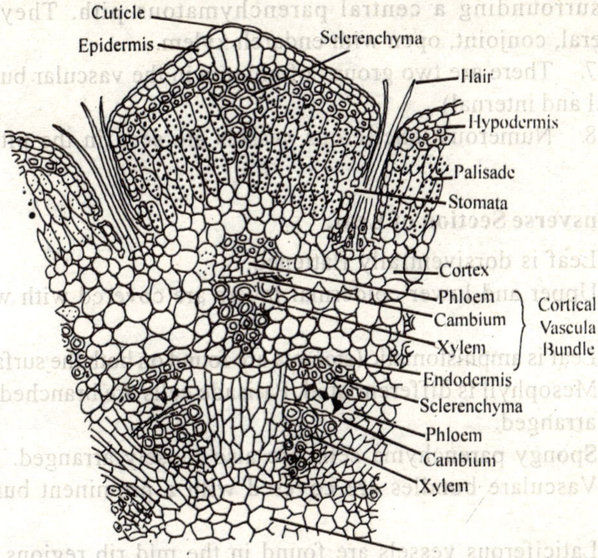

Fig 4.28 Casuarina, T.S. of Stem (A part enlarged)

4. Leaves opposite and decussate. Inflorescence is a number.

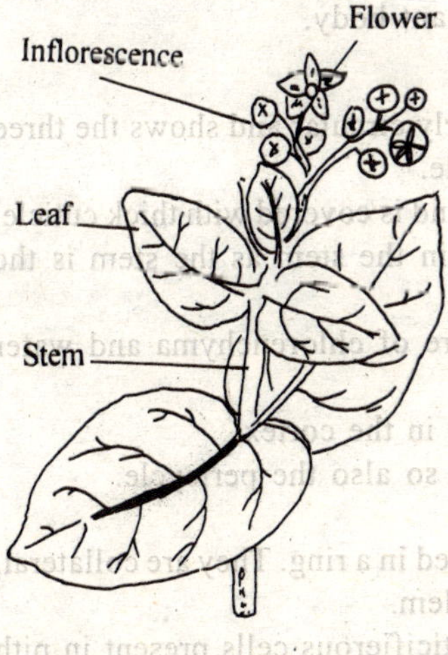

Fig 4.29 Calotropis, Habit

Transverse Section of Stem

1. Transverse Section shows epidermis, cortex and stele.
2. Epidermis is single layered. Thick cuticle and waxy deposition covers the epidermis.
3. Epidermal hairs present.
4. Cortex is divided into two zones - outer collenchymatous and inner parenchymatous.
5. Endodermis surrounds the stele. Pericycle is represented by patches of sclerenchyma.

6. Vascular bundles are many in number arranged in the form of a ring surrounding a central parenchymatous pith. They are bicollateral, conjoint, open with endarch xylem.

7. There are two groups of phloem in the vascular bundle (external and internal).

8. Numerous laticiferous vessels are seen in the pith.

Transverse Section of Leaf.

1. Leaf is dorsiventrally flattened.
2. Upper and lower epidermal layers are covered with waxy material.
3. Leaf is amphistomatic (stomata are found on both the surface).
4. Mesophyll is differentiated. Palisade cells are branched and loosely arranged.
5. Spongy parenchyma cells are also loosely arranged.
6. Vasculare bundles are covered with a prominent bundle sheath.
7. Laticiferous vessels are found in the mid rib regions.

Euphorbia tirucalli

Morphology

1. The plant body is a shrub.
2. Leaves are small in size and deciduous.
3. Stem performs the function of photosynthesis.
4. Milky latex is present in the plant body.

Transverse Section of Stem

1. A transverse section is nearly circular and shows the three regions - epidermis, cortex and stele.
2. Epidermis is single layered and is covered with thick cuticle.
3. Sunken stomata are found in the stem as the stem is the chief photosynthetic organ.
4. Cortex consists of a mixture of chlorenchyma and water storing parenchyma.
5. Latex cells are also present in the cortex.
6. Endodermis is not distinct, so also the pericycle.
7. Vascualr, bundles are arranged in a ring. They are collateral, conjoint and open with endarch xylem.
8. Pith is parenchymatous. Laticifierous cells present in pith also.

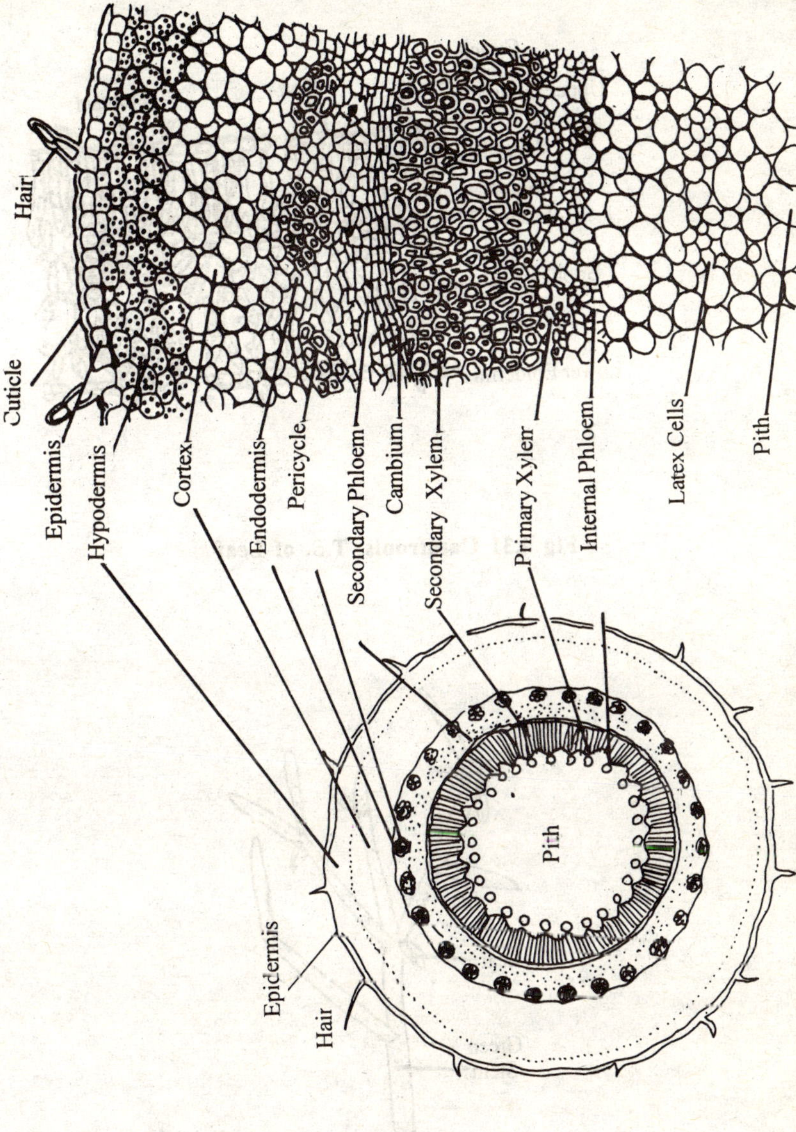

Fig 4.30 Calotropis, T.S. of Stem
A. Ground Plan
B. A part enlarged

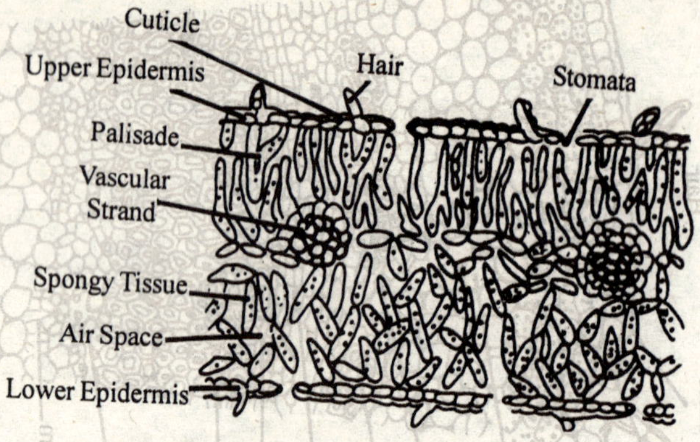

Fig 4.31 Catotropis, T.S. of Leaf

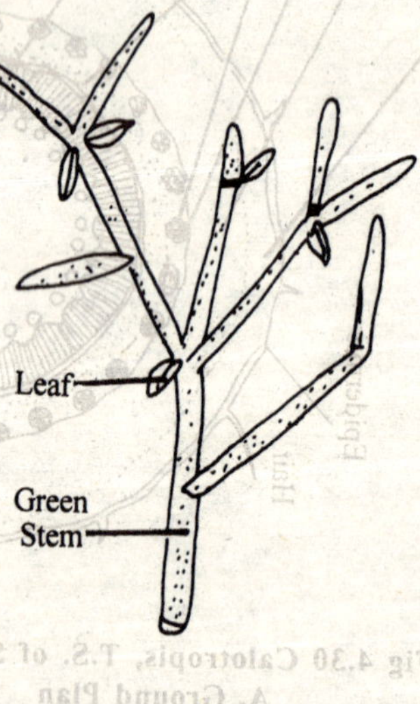

Fig 4.32 Euphorbia tirucalli, Habit

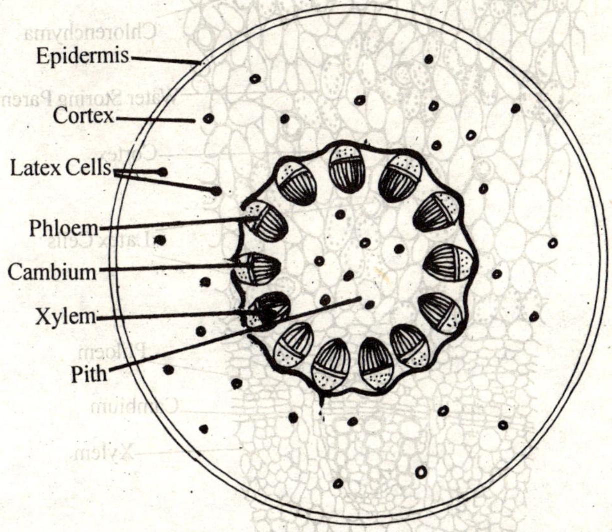

Fig 4.33 Euphorbia tirucalli, T.S. of Stem (Ground Plan)

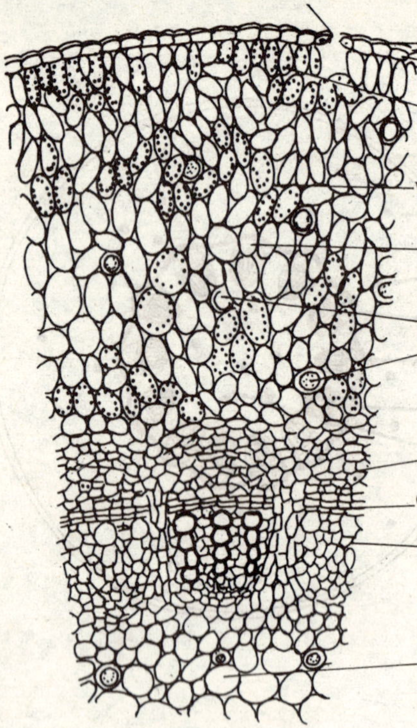

Fig 4.34 Euphorbia tirucalli, T.S. of Stem (A part enlarged)

Asparagus

Morphology

1. Plant body is a herb. Stem is cylindrical.
2. At the nodal region leaves and branches arise.
3. Leaves are reduced to scales.
4. In the axil of the scale leaves green flattened branches of one internodal length arise in clusters. These look like leaves. In reality they are modified branches know as cladodes.
5. As the leaves are reduced, cladodes take up the function of photosynthesis.
6. Roots are fasciculated, they are swollen into white tuberous structures. They store food material.

Transverse Section of Stem

1. A transverse section has a wavy outline due to the presence of ridges.
2. Outermost layer is epidermis which is single layered. A cuticle is deposited outside the epidermis.
3. Next to the epidermis is found a few layers of chlorenchyma constituting the hypodermis
4. Internal to the hypodermis is the endodermis.
5. Pericycle is represented by a few layers of sclerenchyma.
6. Next to the pericycle is the ground tissue made upf of parenchyma cells with intercellular spaces.
7. Vascular bundles are embedded in this ground tissue. The bundles towards the central region are larger than those towards the periphery.

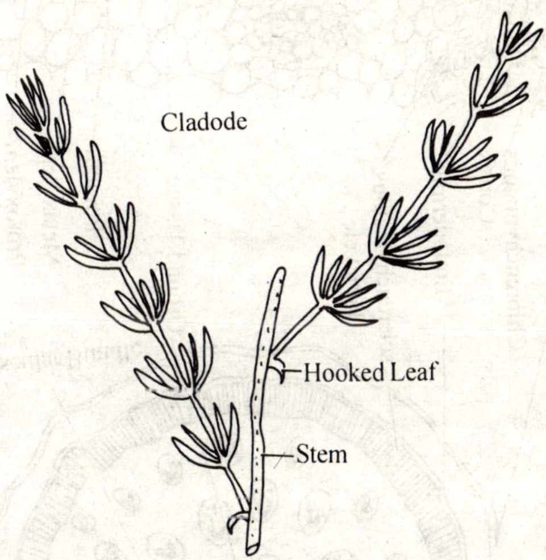

Fig 4.35 Asparagus, Habit

8. Vascular bundles lack a bundle sheath. They are conjoint, collateral and closed with endarch xylem. Xylem is 'V' shaped.

Transverse Section of Cladode

1. The cladode is bifacial (dorsiventrally flattened).
2. Upper and lower epidermis each are single layered.
3. Towards the margins the two epidermal layers meet as the mesophyll becomes thin
4. Stomata are few in number

5. Mesophyll is differentiated, but the palisade is found all around the cladode internal to both upper and lower epidermis.

6. Internal to the palisade tissue is spongy parenchyma with many intercellular spaces.

7. There are two vascular bundles in the centre surrounded by thin walled closely packed parenchyma cells.

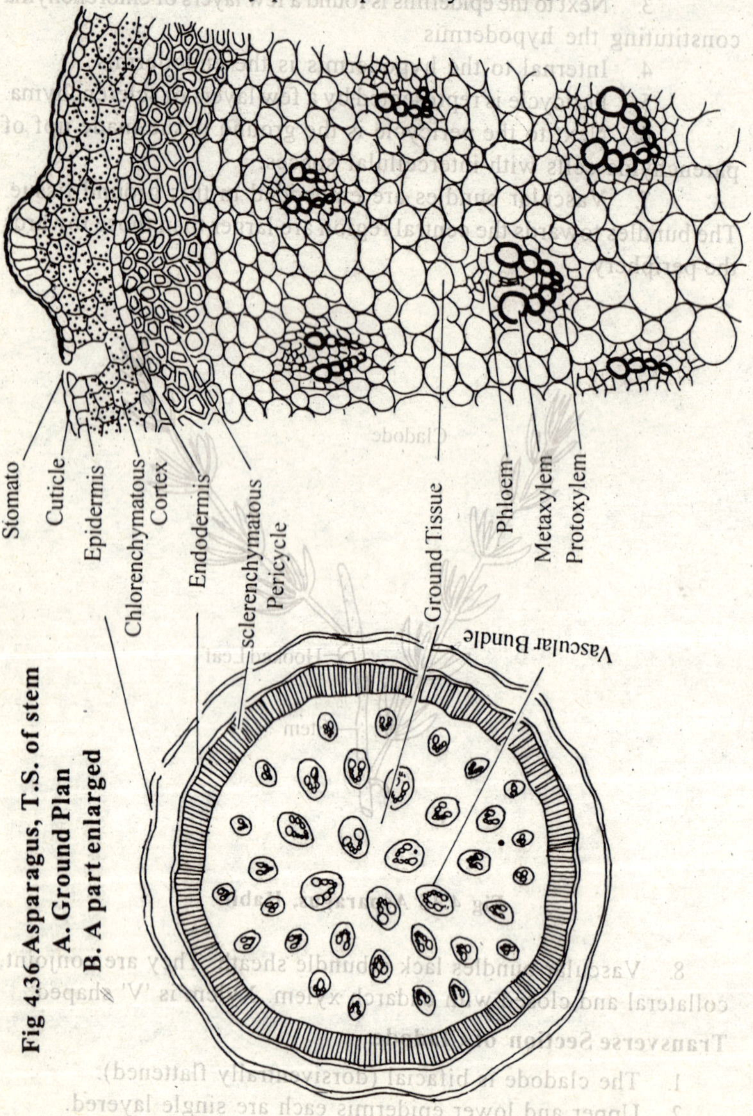

Fig 4.36 Asparagus, T.S. of stem
A. Ground Plan
B. A part enlarged

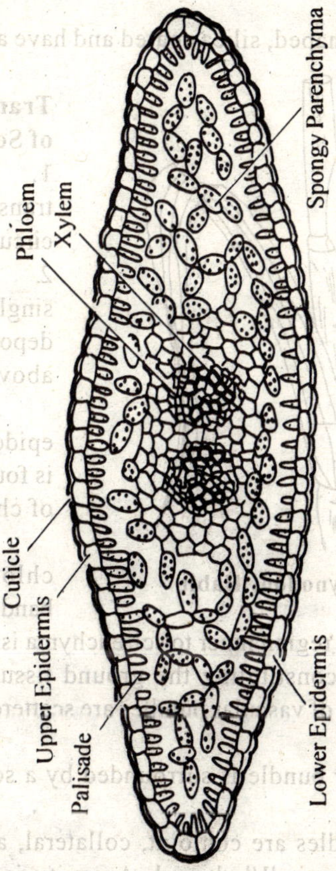

Fig 4.37 Asparagus, T.S. of Cladode

CYNODON

Morphology :

1. It is commonly called doob grass
2. The plant usually grows in open fields, lawns etc.
3. The stem is a horizontal runner from which aerial shoots are produced.
4. Leaves are ribbon shaped, silica coated and have a long sheathing leaf base.

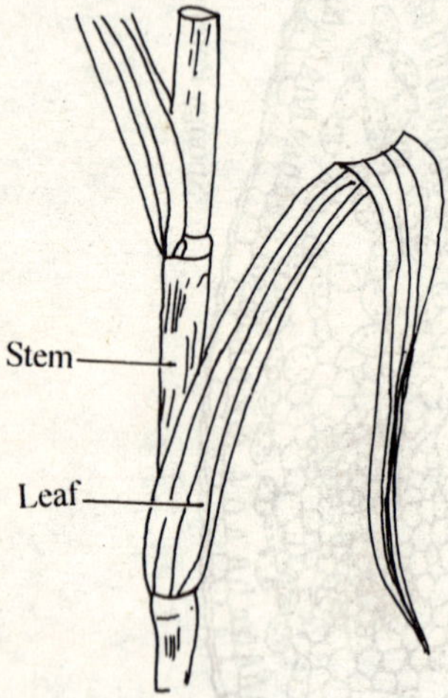

Fig 4.38 Cynodon, Habit

Transverse Section of Setem

1. Outline of transverse section is circular
2. Epidermisis single layered with a deposition of cuticle above.
3. Next to the epidermis internally is found a few layers of chlorenchyma.
4. Next to the chlorenchyma is a band of sclrenchyma.
5. The entire region inner to sclrenchyma is filled by a mass of parenchyma cells constituting the ground tissue.
6. A number of vascular bundles are scattered in the ground tissue.
7. Each vascular bundle is surrounded by a sclrenchymatous bundle sheath.
8. Vascular bundles are conjoint, collateral, and closed with endarch xylem. Xylem is 'V' shaped. A cavity is formed by the dissolution of protoxylem elements. Phloem is situated above the xylem.

Transverse Section of Leaf

1. Leaf is isobilateral.
2. Upper and lower epidermal cells are single layered and surrounded by cuticle.
3. Stomata are found in the upper epidermis.
4. There are certain special cells in the upper epidermis called motor cells or bulliform cells. These cells help the leaf to roll up during dry conditions

5. Mesophyll is undifferentiated Chlorenchyma cells with little or no intercellular speces constitute the mesophyll tissue.

6. A number of vascular bundles are arranged paralleyly (venataion is parallel). Xylem and phloem are well developed. Occasionally pads of sclerotic cells abutt the vascular bundles both above and below.

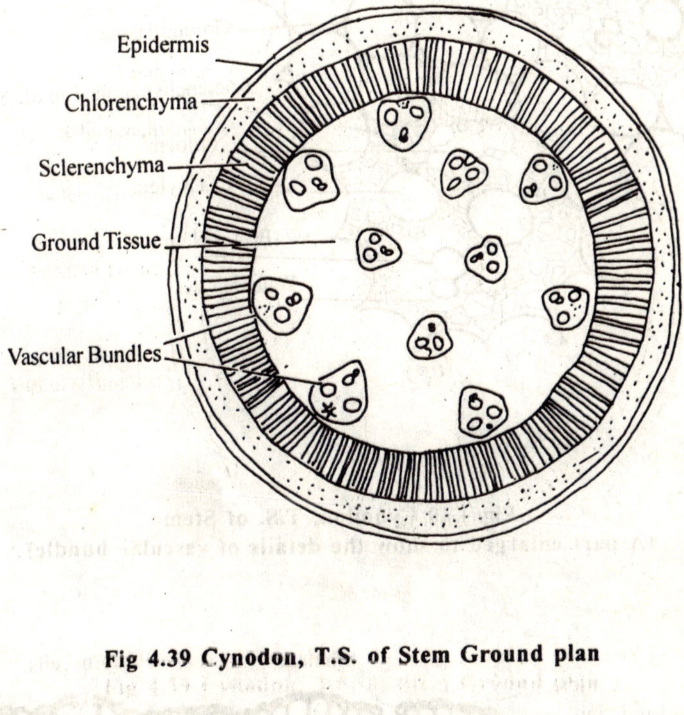

Fig 4.39 Cynodon, T.S. of Stem Ground plan

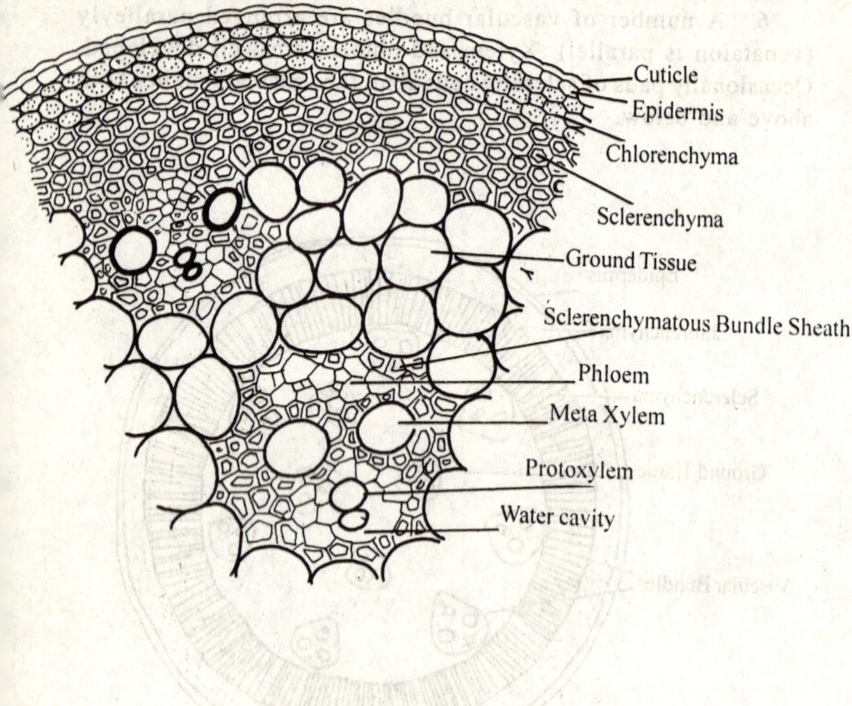

Fig 4.40 Cynodon, T.S. of Stem
(A part enlarged to show the details of vascular bundle)

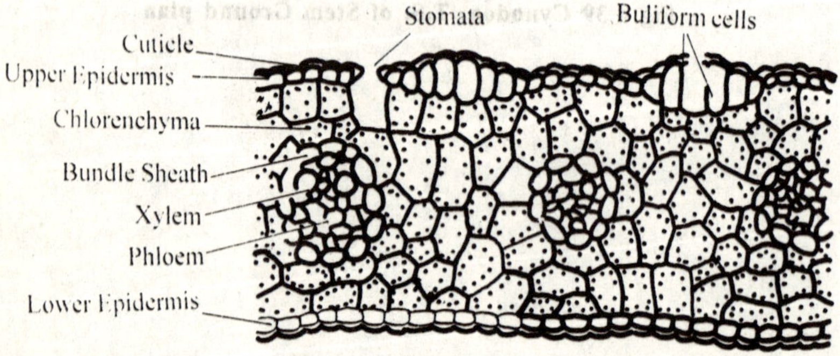

Fig 11.46 Cynodon, T.S. of Leaf

MUEHLENBECKIA

Morphology

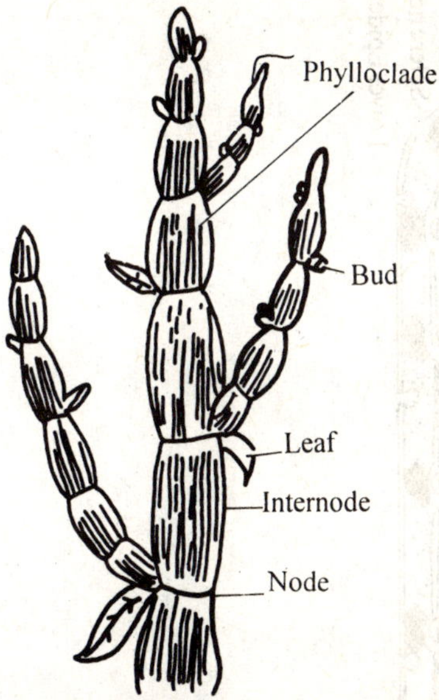

Fig 4.41 Machlenbeckia, Habit

1. Plant is a xerophytic shrub.
2. The entire stem is modified into a green flattened ribbion like structure called phylloclade.
3. Leaves are absent normally.
4. Occasionally small leaves arise from the nodal regions of the phylloclade
5. Stem takes up the function of photosynthesis
6. Flowers are borne in clusters at the nodal region.

Transverse Section of Phylloclade

1. The phylloclade is dorsiventally flattened.
2. Upper and lower epidermis are made up of a single layers of cell each. Cuticle is found above the epidermis.
3. Internal to the epidermis all round there are patches of sclrenchyma.
4. In between the sclcrotic patches chlorenchyma cells are found.
5. Endodermis is irregualr. Cells have starch grains. Pericycle is internal to the endodermis.
6. Vascular bundles are many in number; they are collateral open and endarch.
7. There is a parenchymatous pith in the centre.

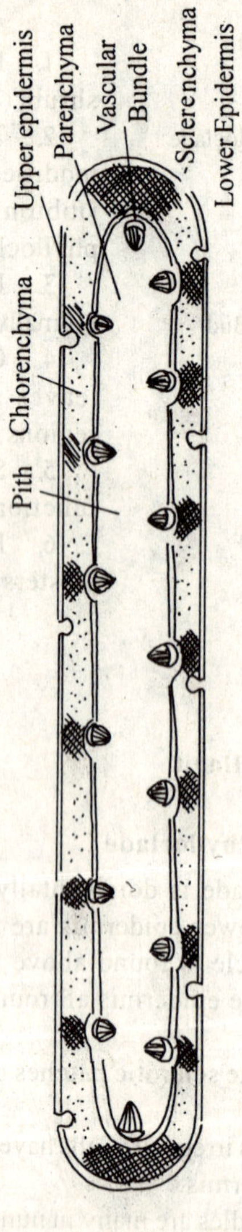

Fig 4.42 Muehlenbeckia,

T.S. of Phylloclade (Ground Plan)

RUSCUS

Morphology

1. The plant is a tall herb or shrub.
2. Stem is cylindrical hard and green.
3. Branches are modified into the green flattened leaf like structures called plylloclades. Phyllocladeus are two internodes long. At the middle of the phylloclade (node) a small inflorescence arises proving the stem nature of the phylloclade.
4. Leaves are reduced to scales.

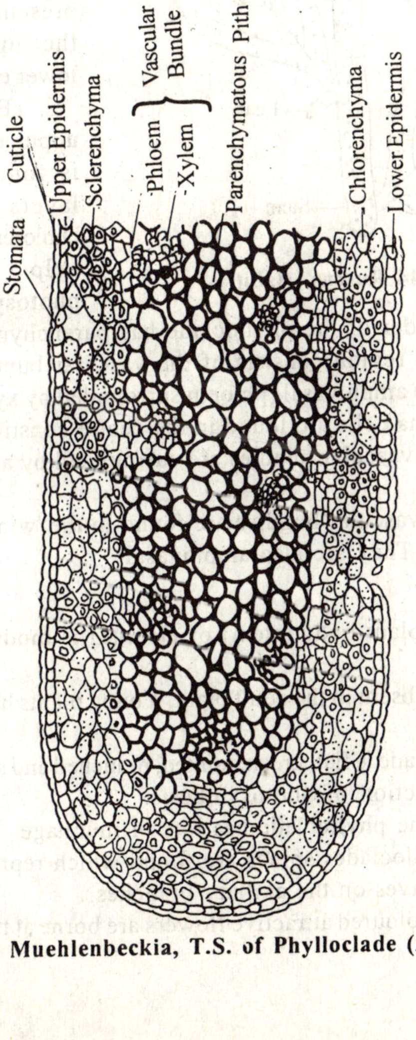

Fig 4.43 Muehlenbeckia, T.S. of Phylloclade (A part enlarged)

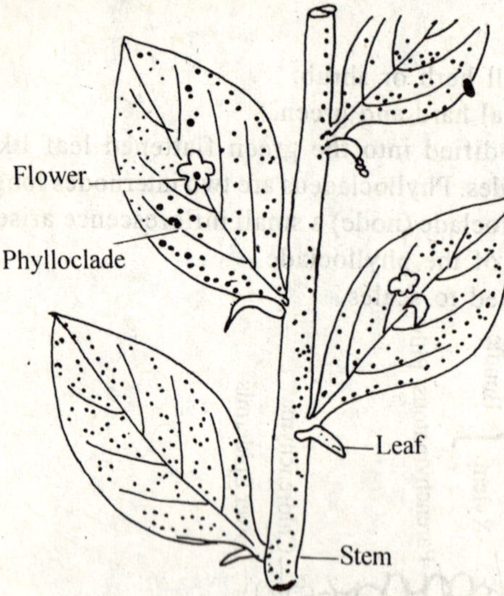

Fig 4.44 Ruscus, Habit

Transverse Section of Phylloclade

1. Epidermis is the outermost layer. It is covered by a deposition of the cuticle.

2. Stomata are present both on the upper and lower epidermis.

3. Below the upper epidermis is found a few layers of chblorenchyma to help in photosynthesis.

4. The remainder of the phylloclade has parenchyma tissue.

5. The central region consists of the vascular bundles. The vascular bundles are amphivasal (phloem surrounded by xylem) and group of parenchyma cells are found in the centre constituting the pith region. All the vascular bundles are surrounded by a common bundle sheath.

6. Additional vascular bundles are found in the 'wing' region of the phylloclade. They are also amphivasal.

Opuntia

1. The entire plant body (stem) of *Opuntia* is modified into a phylloclade.

2. Leaves are absent in the adult plant. Young plants have small caducous leaves.

3. The phylloclade is flat, green, jointed, branched and succulent. It performs the function of photosynthesis.

4. Internally the phylloclade has lots of mucilage

5. On the phylloclades spines are found which represent the modifications of leaves on the axillary branches.

6. Large, red coloured attractive flowers are borne at the nodes.

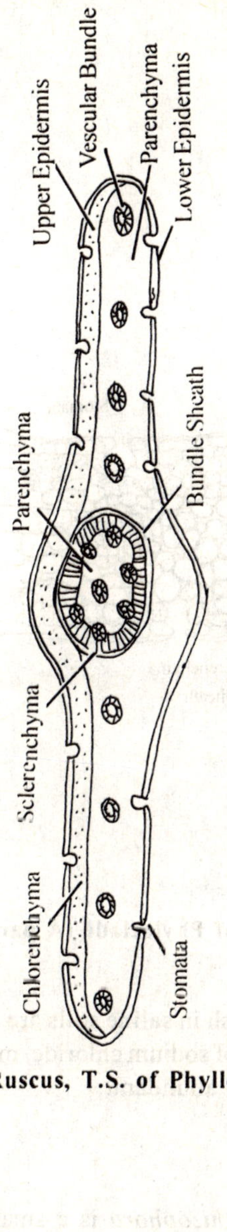

Fig 4.45 Ruscus, T.S. of Phylloclade (Ground Plan)

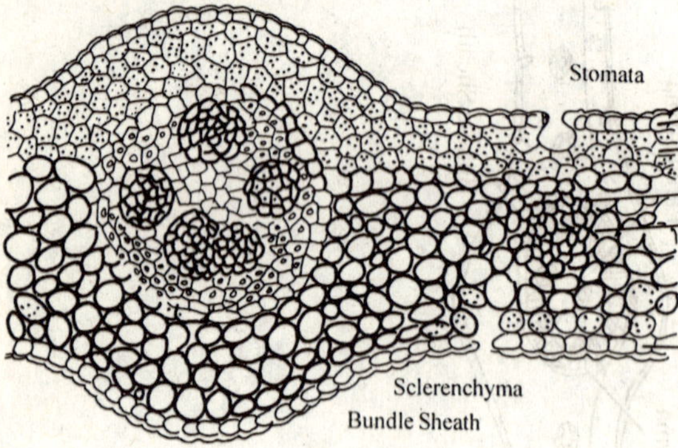

Fig 4.46 Ruscus T.S. of Phyloclade (A part enlarged)

HALOPHYTES

Plants which grow and flourish in saline soils are called halophytes. In saline or marshy soil, salts of sodium chloride, magnesium sulphate and magnesium chloride are abundant.

Rhizophora

Morphology

1. The plant body of *Rhizophora* is a small tree growing in the backwaters of sea.
2. Stem is anchored to the loose soil with the help of stilt roots.
3. In addition to the stilt roots certain special kind of roots grow apogeotropically and come above the water level. These are

called pneumatophores or breathing roots. The pneumatophores have small pores called air pores which absorb air and provide the roots below as the soil has poor aeration due to water logging.

4. The seeds find it difficult to germinate in saline water due to osmotic imbalance. In order to overcome this the seeds germinate *in situ* i.e., they develop into small seedlings while still attached to the parent plant. This phenomenon is called Vivipary.

5. The seedlings get detached from the plant and fall straight into water and get dispersed due to water current.

6. Halophytes exhibit many xerophytic features such as, bushy growth, thick leathery leaves, hairy outgrowths, sunken stomata etc. This is due to physiological dryness.

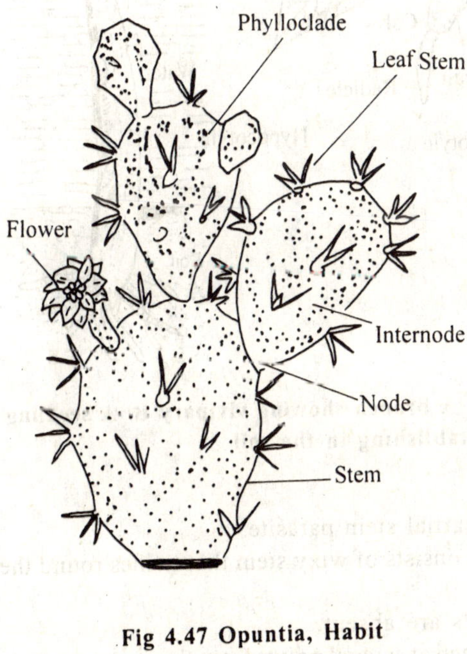

Fig 4.47 Opuntia, Habit

PARASITES.

Parasitic plants are achlorophyllous. They depend on green plants for their nutrition. Parasities are of two types - partial and total. Partial parasites depend on the host only for some of their requirements while total parasites are wholly dependent on the host plant. A few examples are discussed below.

Viscum album

1. This is a small herb a partial parasite growing on the branches of trees.

2. The stem is jointed. Leaves arise at the nodal region.

3. The stem produces haustoria which penetrate the host stem and absorb nutrition from the stem.

4. Seeds are sticky and when birds leave it on twigs (after eating the fruit), they (seeds) germinate there with the radicle coming out as a sucker

5. The sucker after penetrating the host tissue gives out branches which form an internal absorbing system.

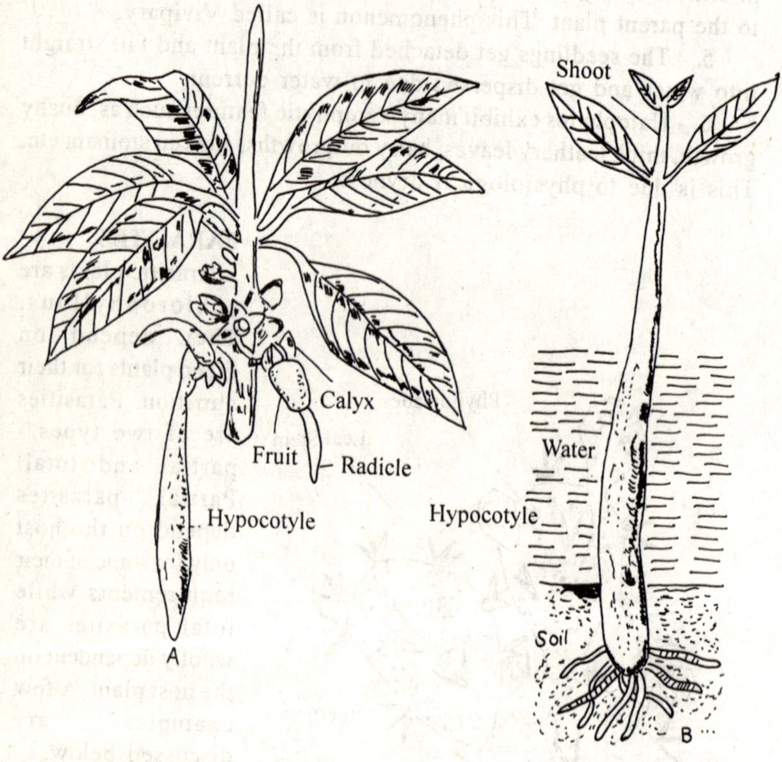

Fig 4.48 Rhizophora A. A branch showing vivipary B. A seedling Establishing in the soil

Cassytha filiformis

1. The plant is a partial stem parasite.
2. The plant body consists of wixy stem that twines round the host stem.
3. Leaves and roots are absent.
4. Haustoria produced at several points from the stem penetrate the host stem and absorb nutrition.
5. The stem is greenish in colour and hence it can carry out some amount of photosynthesis.

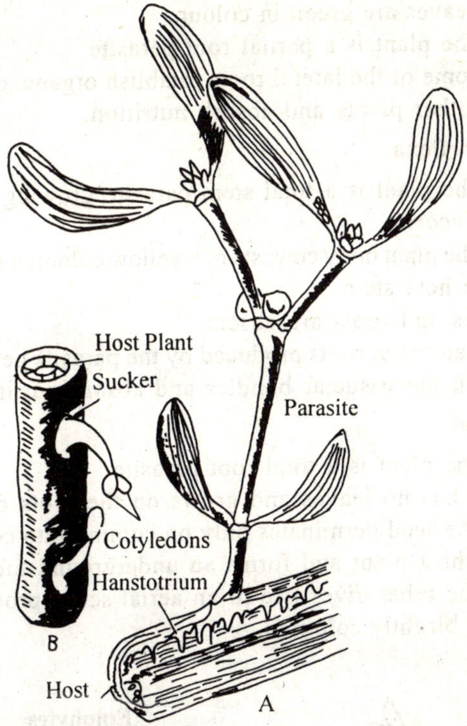

Fig 4.49 Viscum, A. Habit
B. A Stages of Germination on host stem.

Fig 4.50Cassytha filformisA partial Stem parasite (Habit)

Santalum album

1. This is a famouxs oil yielding tree.
2. Leaves are green in colour.
3. The plant is a partial root parasite.
4. Some of the lateral roots establish organic contact with the roots of other plants and absorb nutrition.

Cuscuta reflexa

1. The plant is a total stem parasite growing on host plants such as *Tecoma*.
2. The plant body consists of a yellow coloured stem that twines round the host stem.
3. Leaves and roots are absent.
4. Haustorial roots produced by the parasite penetrate the host stem reach the vasuclar bundles and absorb nutrition.

Orobanche

1. The plant is a total root parasite.
2. It has no leaves and grows on the roots of host plants.
3. The seed germinates only on coming into contact with the root of a host plant and forms an underground tuber.
4. The tuber gives rise to an aerial scape producing flowers which are birghtly coloured.

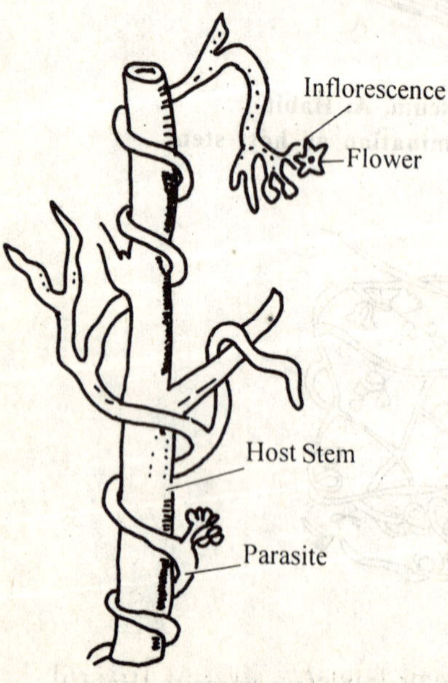

Fig 4.51 Cuscuta - A total stem Parasite, Habit

Epiphytes

Epiphytes grow on other plants only for shelter and sunlight. They are also called aerophytes. A number of plants belonging to families orchidaceae, Araceae, Cactaceae are epiphytic. *Vanda* one of the common orchid is discussed here.

Vanda roxburghii

1. The plant is a small herb growing on the branches of trees.
2. There are two types of roots - clinging roots and aerial roots (epiphytic roots).

3. Clinging roots anchor the epiphytes to the branch of the host plant.

4. Aerial roots hang down in air and absorb atmospheric moisture with the help of a tissue called velamen. This is the only source of water for the epiphyte as it does not absorb any water or nutrition from the host.

5. The epiphytic roots are green in colour and can carry out photosynthesis. Hence they are also called assimilatory roots.

6. Epiphytes develop many xeromorphic features as they get very little water. These are stunted growth, leathery leaves, thick cuticle, sunken stomata etc.

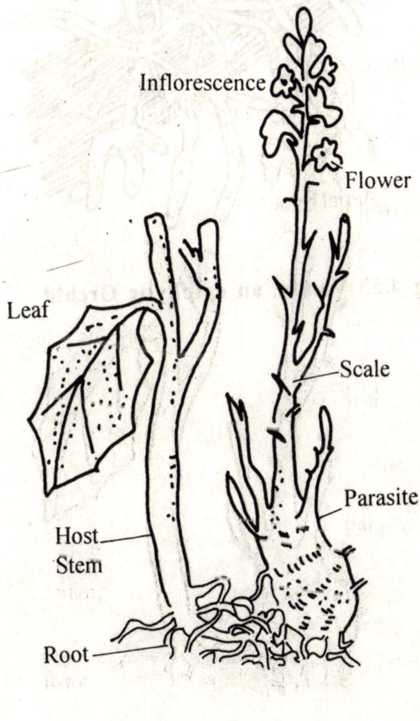

Fig 4.52 Orobanche, A total root parasite, Habit

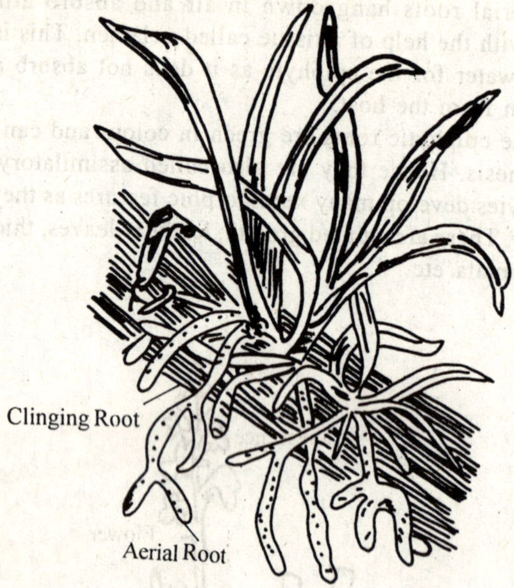

Fig 4.53 Vanda, an epiphytic Orchid

Part - Two
Plant Physiology

5

PLANT CELL

1. *Protoplasmic Streaming*

Materials : Leaves of *Elodea*, staminal hairs of *Tradescantia*, slides, cover slips, water, compound microscope.

Procedure : Collect fresh leaves or dissect out the staminal hairs. Place them on the slide and cover with the help of cover slip adding a drop of water. Observe under high power of the microscope, specially the chloroplasts in *Elodea* leaf.

Observation : The chloroplasts appear to move in a circulatory fashion around the vacuole.

In the case of *Elodea* there is a large central vacuole and the chlorplasts move in a clockwise or anticolock wise direction. This is known as Rotation. In the staminal hairs of *Tradescantia* there are many vacuoles and the contests move around each vacuole independently and it is known as Circulation.

Inference : The movement of the chloroplasts seen under the microscope is actually due to the circulation or streaming of protoplasm. The movement probably occurs due to the sol-gel transformation of protoplasm. Microtubules of cytoplasm may also contribute to the movement.

2. *Importance of Living Membrane*

Materials: Beet root, cork borer, water, ice, spirit lamp, beakers, test tubes, thermometer, alcohol, water bath, test tube stand.

Procedure : Cut small cylinders of (approximately one c.m length) beet root with the help of a cork borer. Place the cylinders in a beaker and wash throughly under running water (to remove the pigments that ooze out) till the exudation of colour stops.

Fill seven test tubes with about 20ml of water. to each test tube add 2 cylinders and keep the test tubes in water bath separately at different temperatures (25°, 30°, 40°, 50°, 60°, 75° and boiling temperature) for about 5 minutes. Also place two cylinders in ice cold water (20 ml) contained in a test tube for 5 minutes. Repeat the procedure with absolute alcohol and acetone.

Observation : Pigments from the cylinders of beet root start leaching out and the intesity of colour increases with increase in temperature with maximum being in boiling water. Increased pigment leaching is noticed in organic solvents like alcohol and acetone.

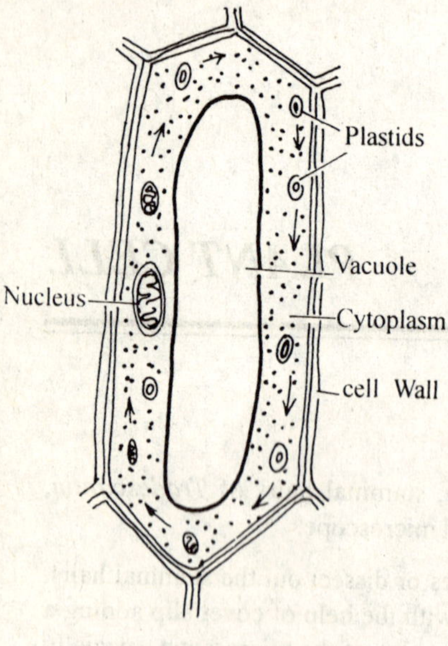

Fig 5.1 Streaming of Protoplasm in Hydrilla (Leaf cell)

Inference : Under Normal temperature, membranes do not allow pigments to leach out, with increased temperature however permeability increases and in boiling water the membrane retentivity is lost resulting in maximum leaching. Alcohol and acetone treatments also result in membrane losing retentivity.

6

SOLUTIONS, SUSPENSIONS AND COLLOIDS

1. Preparation of 5% solution of sucrose (w/w)

Materials : Sucrose, water, measuring cylinder, Balance etc.

Procedure : Weigh 5 gms of analytical grade sucrose and put it into a 100 ml volumetric flask. Add little water and stir to dissolve sucrose. Now make up the volume to 100 ml.

Conclusion : A true solution of sucrose is obtained by this procedure as the solute completely dissolves in water. This is a percentage solution weight by weight (w/w) i.e 5 gms sucrose in 95 gms of water

2. Preparation of 5% solution of Sucrose (wl)

Materials : Same as in experiment No. 3.

Procedure : Weight 5 gms of analytical grade sucrose and dissolve it in 100 ml of water. Stir until sucrose completely dissolved.

Conclusion : This is a percentage solution weight by volume (w/v) i.e., 5 gms of sucrose in 100 ml of water.

3. Preparation of 80% alcohol (v/v)

Materials : Absolute alcohol, distilled (water, measuring cylinder, beaker etc.)

Procedure : Take 80 ml of absolute alcohol and to it add 20 ml of distilled water. The volume will now be 100 ml.

Conclusion : This is a percentage solution of (liquid in liquid) volume by volume. The solution is 100 ml of 80% alcohol.

4. *Preparation of 1 Molar solution of Glucose*

Materials : Sucrose, distilled water, balance, measuring cylinder etc.

Procedure : Weigh 180.16 grams of glucose and dissolve it in some quantity of water. Then add water to make up the final volume to 1 litre

Conclusion : 1 molar solution of glucose will have 1 mole of glucose in one litre of water. What is a mole? Gram molecular weight of a substance dissolved in one litre of water will give one mole for every litre. What is gram molecular weight? Weight in grams of a substance equivalent to its atomic weight. In the case of glucose the atomic weight is $C_6 H_{12} O_6$ = 12 x 6 + 12.6 + 16.6 = 180.16x i.e., 180.16 gms of glucose in 1 gram molecular weight. Hence the above solution is 1 molar (1m). To prepare a 2 molar solution dissolve 360.32 gms of glucose in distilled water and make up the volume to one litre.

5. *Preparation of 1 molal solution of glucose*

Materials : Same as in experiment no - 6

Procedure : Dissolve a gram molecular weight of glucose (180.16 gms) in one litre of water.

Conclusion : This is slightly different from a molar solution in that the final volume of the solution will be more than one litre. Here one litre of a solution has slightly less than 1 mole of glucose per litre of the solution.

6. *Preparation of 1 normal solution of NaoH*

Materials : NaoH, Distilled water, balance, measuring cylinder etc.

Procedure : Weigh 40 gms of NaoH and dissolve it in some quantity of water. Make up the final volume to one litre.

Conclusion : In a normal solution a gram equivalent of a substance is dissolved in water to finally make one litre of the solution. What is gram equivalent weight? Gram molecular weight divided by the hydrogen equivalent. In NaoH the total molecular weight is 23 + 17 = 40 gms. Since there is only one hydrogen its value is 1. Hence 40 gms divided by 1 is 40. The gram equivalent weight of NaoH is 40. In this case gram molecular weight and gram equivalent weight are same as there is only one hydrogn.

7. *Preparation of 1 normal H_2So_4*

Materials : Commercial, H_2So_4, Water, Measuring cylinder etc.

Procedure : Dissolve 49 gms or 28 ml H_2So_4 in one litre of water.

Conclusion : 49 gms is the gram equivalent weight of H_2So_4. Since it cannot be weighed it can be converted to volume by the formula V=M/D. Specific gravity of H_2So_4 is 1.84. Hence 49/1.8 = 26 ml. Since H_2So_4 is 8% pure about 28ml of H2SO4 is required to prepare 1 litre of 1N solution of H_2So_4.

8. Preparation of a Suspension

Materials : Barium Chloride Solution, test tubes, dilute H_2SO_4, Test tube stand etc.

Procedure : Pour a little quantity of $BaCl_2$ solution in a test tube. Add a few ml of sulphuric acid. Shake the mixture thoroughly.

Observation : A white precipitate of barium sulphate settles down.

Conclusion: In a suspension, particles remain dispersed in the medium. They do not dissolve. Suspensions are generally unstable and gradually the particles settle down and the pure solvent can be separated.

9. Preparation of an Emulsion

Materials : Groundnut oil or mustard oil, water, test tube etc.

Procedure : Take little amount of oil (10ml) in a test tube. Add 20 ml of water and shake vigorously.

Observation : Oil and water are imiscible. Water is polar while oil is non polar, further the density of oil being less than that of water, it (oil) floats on the surface of water. . Thus two layers (oil and water) are formed in this emulsion.

10. Demonstration of Brownian Movement

Materials : Microscope, latex of *Calotropis*, or *Nerium* India ink, slide, cover slip etc.

Procedure : Place a drop of the latex on the slide and cover it with cover slip. Observe it under the high power objective of the microscope

Dilute India ink abou t10 times and follow the same procedure as for latex.

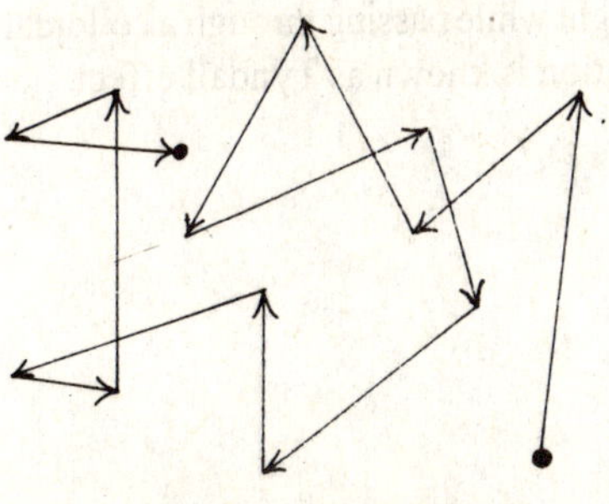

Fig 6.1 Demonstration of Brownian Movement

Observation : Small particles in the liquid get illuminated and move to and fro in the microscopic field.

Conclusion : The particles are in constant motion due to their inherent kinetic energy in a colloidal system. Uneven bombardment of the particles by the dispersion medium results in this movement. This movement was first discovered by Brown, hence it is called Brownian movement.

11. Demonstration of Tyndall effect

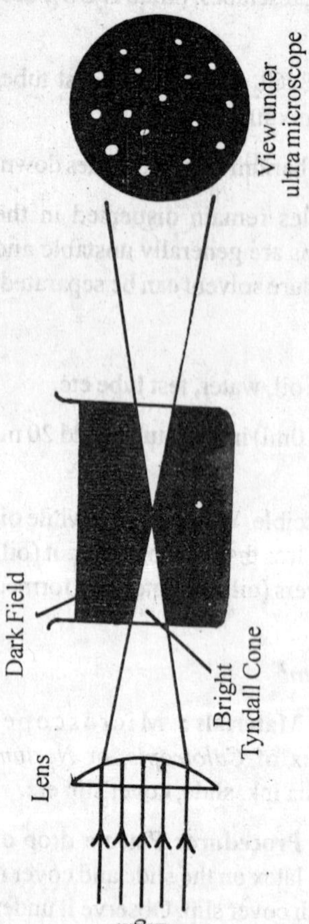

Fig 6.2 Demonstration of Tyndall effect

Materials : 1% glucose solution, fruit juice, beaker, pencil, torch, water.

Procedure : Take two beakers and pour glucose solution in one and fruit juice in the other. Keep the beakers in darkness. Direct the torch towards each beaker and switch on the light so that the beam of light passes through the solution or fruit juice. Observe the behaviour of the light beam in both the cases.

Observation : The beam of light will be normal in the case of beaker having glucose solution., while in the case of fruit juice the light beam gets scattered.

Conclusion : Fruit juice being a colloideal solution the light beam gets scattered as the particles are in constant motion. This phenomenon of scattering of light while passing through a colloidal solution is known as Tyndall effect.

MOVEMENT OF SUBSTANCES

1. *Separation of Starch and Salt by dialysis*

Materials : Collodion bag, starch, Nacl, water, test tube, spirit lamp, silver nitrate, iodine, thread, stand, glass rod, beakers etc.

Procedure : Prepare 2 % solution of Nacl and a colloidal solution of starch (Add 1 gm of starch to 50 ml of boiling water with constant stirring). Mix the two solutions in equal volumes. Fill the mixture in a dialysis bag (collodion bag) and immerse it as shown in the figure in a beaker containing distilled water. After about 1 or 1 1/2 hours, test the water of the beaker with silver nitrate and Iodine Solutions separately.

Observation : On treated with silver nitrate, the water turns milky showing the presence of salt. It does not turn blue idicating starch has not diffused out of the dialysis bag.

Conclusion : The collodion membrane is differentially permeable, while it allows the salts to go out, it does not allow starch to go out.

Movement of Substances

2. *Demonstration of Diffusion :*

Materials : Beaker, water, $CuSo_4$, crystals.

Procedure : Place few crystals of copper sulphate in a beaker containing water and observe.

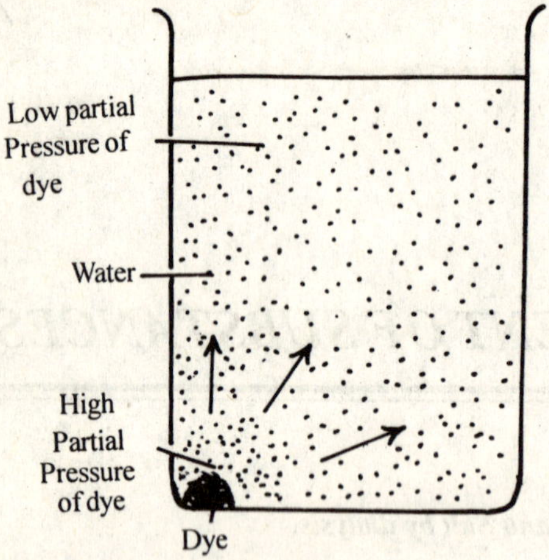

Fig 7.1 Demonstration of Diffusion

Observation :

The crystals dissolve slowly in water. The particles may be seen moving away rapidly from the crystal to other parts in water. After the crystals completely dissolve this unidirectional movement stops and there will be to and fro movement.

Conclusion :

The above experiment demonstrates the diffusion of solid particles. Diffusion is defined as the particles of solid, liquied organs distributing themselves evenly throughout the available place.

3. *Demonstration of Imbibition*

Materials : Dry seeds, water, pertridish.

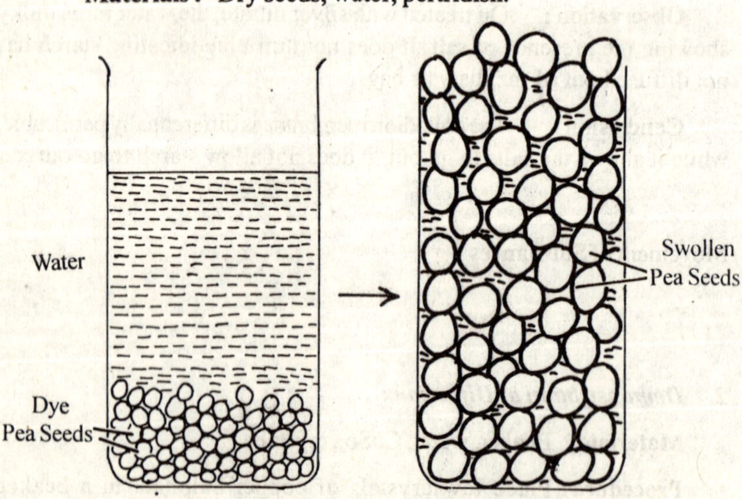

Fig 7.2 Demonstration of Imbition

Procedure : Soak the seeds in water overnight. Observe the ssize of the seeds. Check the termperature of water.

Observation : The seeds become swollen. After some time the seed coat breaks open. There will be slight rise in termperature of water.

Conclusions : Seeds swell due to imbition of water. The cotyledons swell at a much faster rate than seed coat, the seed coat breaks open exposing the cotyledons. This helps in germination. Imbibition involves both *absorption* as well as *adsorption*. Due to adsorption water molecules are held over the surface of the seed. As they cannot move they lose their energy in the form of heat into water; hence the rise in temperature of the water.

4. *Demonstration of Imbition Pressure*

Materials : Imbibitional pressure instrument (disatometer), dry bean seeds, water and weights.

Procedure : Keep the apparatus as shown in the figure. The apparatus consists of a cylindrical jar fitted with a metallic piston. The metallic piston has a flat pan on its top to which weights can be attached. To the upperr end of the piston is atached a pointer (indicator) which glides above and below on a graduated arc scale fitted to a stand.

Keep some dry bean seeds in the apparatus and add water into it. Press the piston over the seeds by attaching weights on the pan. Note down the initial reading on the scale. Leave the experiment for about two hours. Note the final reading on the scale.

Observation : As the seeds soak in water they become swollen and push the piston above. As a result the indicator moves down on the scale certain degrees.

Conclusion : Due to imbibition the seeds swell and as they are confined in a closed container they develop a pressure and as a result the piston moves up. If we put enough weight on the pan so that the piston does not move up, that weight or force equals the imbibitional pressure developed by the seeds.

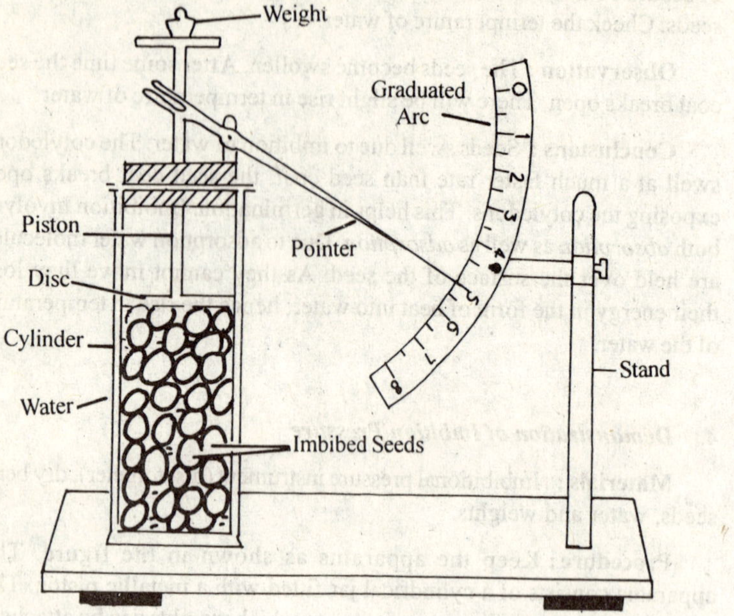

Fig 7.3 Apparatus for measurement of Imbibition Pressure

5. *Physical Demonstration of Osmosis (Thistle Funnel Experiment)*

Materials : Egg membrane, animal bladder or parchment paper, 5% sugar solution, thistle funnel, beaker, water, stand etc.

Procedure : Take a beaker half filled with water. Take a thistle funnel and tie its mouth with egg membrane parchment paper or animal bladder. Fix the thistle funnel to a stand inversley so that the funnel is immersed in water. Fill the funnel with 5% sugar solution. Note the level by making a mark on the stem of the thistle funnel. Allow the experiment to stand for about 30 minutes. Then note the final level of sugar solution. Taste the water in beaker

Observation : The level of sugar solution in the thistle funnel has gone up. Water in the beaker does not taste sweet.

Conclusion : The egg membrane is selectively permeable only to water and not to sugar. Hence, as the solvent concentration is higher in beaker than in the thistle funnel. Water migrates into the thistle funnel and the level of sugar solution rises up. This phenomenon is known as osmosis which may be defined as 'Migration of solvent molecules from the region of *their* higher concentration to the region of *their* lower concentration across a selectively permeable membrane.

6. *Physiological Demonstration of Osmosis (Potato Osmoscope)*

Materials : Potato tubers, 5% sugar solution, water, petridish, pins.

Procedure : Select a large size raw potato and carefully peel off the skin. Cut into two pieces horizontally. Scoop out potato tissues from the centre and fashion a small cup as shown in the figure. Pour a little of sugar solution into the potato cup. Stick a pin inside the cup to note down the level of sugar solution. Keep the potato cup (it is now called a potato osmoscope) in the petridish containing water. Leave the apparatus aside for about 30 minutes. At the end of this period note the final level of sugar solution in the potato osmoscope and stick a pin to mark the level.

Observation : It is observed that the level of the sugar solution has gone up inside the potato cup due to entry of water from petridish.

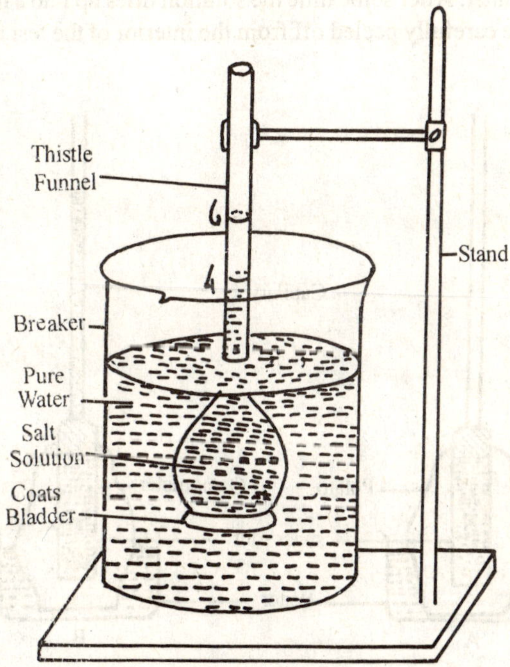

Fig 7.4 Physical Demonstration of Osmosis

(Thistle funnel experiment)

Conclusion : Potato tissues act as a selectively permeable barrier and allow only water but not solute particles to pass through. As a result water which is in higher concentration in the petridish migrates to the potatocup. This migration is due to osmosis. (For definition of osmosis see the previous experiment).

This experiment is known as physiological demonstration of osmosis as it is seen in living tissues, while the thistle funnel experiment in physical demonstration because it is seen in membrane removed out of the living tissues.

7. *Prepartions of a Semi permeable Membrane (Collodion bag)*

Materials : Collodion solution (Cellulose nitrate dissolved in alcohol and ether), water, test tube glass rod, toluol etc.

Procedure : Pour a little collodion solution into a test tube. Keep the test tube in a slanting position and revolve it slowly so that the solution forms a coating on the inner walls of the test tube. Decant the extra solution. Allow it to settle. Then give a second coating of the solution in the same manner. After some time the solution dries up into a membrane which can be carefully peeled off from the interior of the test tube.

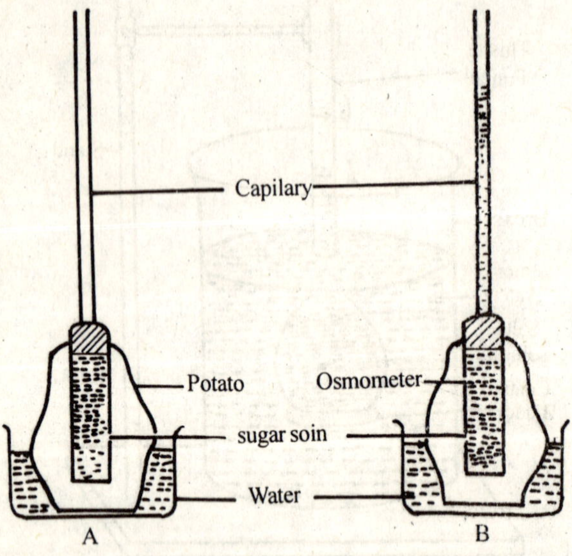

Fig 7.5 Demonstration (Physiological) of Osmosis using Potato Osmoscope

Collodion sac prepared as above can be preserved in water to which a few drops of toluol have been added. This membrane has the property of semipermeability

8. *Demonstration of plasmolysis*

Materials: Onions or Tradescantia leaf, safety blade, sugar solutions of different concerntrations, cover slips, slides, water, microscope etc.

Procedure: Onion peel can be obtained by tearing the onion leaves obliques when a thin transparent layer comes out. Semilarly leaf epidermal peels also can be obtained. Set up a series of slides and place a drop or two of water, 0.25 m, 0.5 m, 0.75 m and 1.0 m sugar solution on the slides as shown below.

Slide:	1	2	3	4	5
	Water	0.25m	0.5m	0.75m	1.0m

(Sugar Solution)

on each slide place a piece of the onion peel and put the cover slip. Observe the shape of the cell, cytoplasm etc and draw a sketch. Leave the preparation aside for 15 minutes. Observe again.

Observation: In slide No. 1 (water) the cells retain their original size and shape and in slide 2 (0.25m) there is slight shrinking of the cytoplasm. A small space will be left between the cell wall and plasma membrane. In slide No. 3 (0.5m) the cytoplasmic shrinkage is considerable and the plasma membrane becomes separate from the cell wall. In slides 4 and 5, cytoplasm completely loses contact with the cell wall and shrinks to its minimum size. The cytoplasm will be either concave or convex in shape.

Conclusion: In ordinary water, the cell retains its shape and size or slight increate might be there due to entry of water. Water is thus hypotonic in comparison with cell sap. In slides 2, 3, 4 and 5 the solution will be increasingly hypertonic as the shrinkage in cytoplasm becomes more due to exit of water. This exit of water from a living cell when placed in a hypertonic solution is known as plasmolysis.

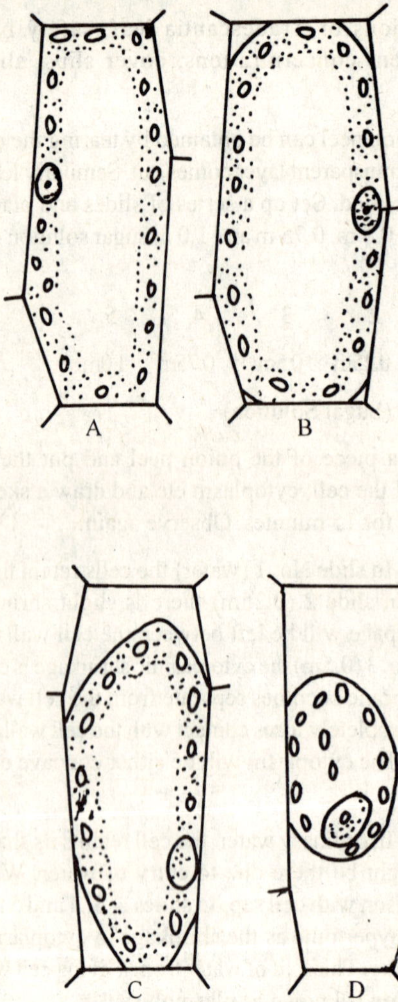

Fig 7.6 Demonstration of Plasmolysis

9. Determination of Osmotic Concentration of Cell sap by Plasmolytic Method

Materials: 1m glucose solution, onion peel, cavity slides, microscope etc.

Procedure: Prepare the following series of dilutions of glucose solutions from 1m glucose solutions.

0.1m, 0.2m, 0.3m, 0.4m, 0.5m, 0.6, 0.7m, 0.8m, 0.9 and 1 m

Take a few drops of the above concentration and place a piece of the onion peel. Observe the shape and size of cells. Leave them aside for 30 minutes and observe.

Observation : Note down as to the increase, decrease or no change of the cell size in different concerntrations of the solution.

Conclusion : The concentration of the solution where there in neither increase or decrease of the onion peel cells equals the concentration of the cell sap. This is the isotonic concentration. Wherever there is increase, it is hypotonic and wherever there is decrease it is hypertonic. For instance if 0.2m is the isotonic concentration osmotic pressure concentration) of cell sap can be calculated as follows.

OP = 22.4 atm x Molarity of external solution x Absolute temperature + Room Temperature

(Approximately 24^0C / 273

OP = 22.4 x 0.2 x 273 + 24 / 273

OP = 4.56

10. Measurement of Diffusion Pressure deficit (Water potential)

Materials : Potato tubers, cork borer, measuring cylinder 1m glucose solution, test tubes, balance, filter paper, scale, test tube stand.

Procedure : Prepare a stock solution of 1m glucose in distilled water (dissolve 180.16 gms in water to make the volume to 1000 ml) and prepare serial dilutions from it as follows.

0.1m, 0.2m, 0.25m, 0.3m, 0.35m, 0.4m and 0.5m. Take 15 ml of each of these dilutions in separate test tubes and keep them in a stand. Take 15 ml of distilled water in another test tube.

Peel the skin of potato and with the help of a cork borer scoop out cylindrical pieces of potato. Cut them into 24 pieces of equal length. Wash them in water and dry with the help of filter paper. Separate the potato cylinders into 8 groups of 3 pieces each. Weigh each group separately and note down the weight. Now put each group into separate test tubes containing glucose solution of different concentrations mentioned. Also place one group into distilled water kept in another test tube. There will be seven dilutions of glucose solution and another will be distilled water. Keep the tubes for about 45 minutes. At the end of this period remove groups (3 pieces each) of potato cylinders from the test tubes, wipe them with filter paper and weigh them separately. Note down the loss, increase or no change in weight in each group. Tabulate the observations as follows

No.	Contents	Initial Weight of potato cylinders	Final Weight of potato cylinder	Difference + or -
1	0.1m	4.5 gms	4.9	0.4+
2	0.2m	4.5 gms	4.7	0.2+
3	0.25m	4.5 gms	4.5	0.0
4	0.3m	4.5 gms	4.4	0.1-
5	0.35m	4.5 gms	4.3	0.2-
6	0.4m	4.5 gms	4.2	0.3-
7	0.5m	4.5 gms	4.1	0.4-
8	Distilled water	4.5 gms	5.0	0.5+

(The weights given above are arbitarary and serve only to illustrrate the example)

Observations : From the above it will be noticed that maximum increase in weight is noticed (0.5 gms) in the case of water, while there is slight increase in 0.1 m and 0.2 m, whereas in molarities above 0.3 there is decrease. In 0.25m however there is neither decrease nor increase. This concentration equals the concentration of the cell sap.

Conclusions : In concentrations below 0.25m there is entry of water into potato cylinders hence their weight increases. In concentrations above 0.2m there is exit of water from the potato cylinders hence their weight decreases. At 0.25m there is neither entry nor exit of water hence no change in weight. This concentration may be used to calculate the osmotic potential (OP) as follows.

OP = CRT

(Where C= molar concentration of the solution, R = gas constant equivalent to 0.082 and T=absolute temperature i.e. 273 + room tmperature Room temperature may be approximately 24°C)

Hence OP = 0.25 x 0.082 x 277

(This value will be a negative value as OP is always calculated as lack of so much pressure to accomodate certain quantity of water to bring its value to zero. When maximum water is present the pressure is zero).

11. Comparison of concentrations of two solutions in relation to cell sap by turrgidity method

Materials : Fresh stem of any herbaceous plant such as *Synedrella*, *Amaranthus*, *Gynura* etc, Salt solution 1(weak) and 2 (strong), watch glasses.

Procedure : Cut the stem (internode) of the herbaceous plant into a cylinders of about 4 cms. Split into two halves logitudinally. Drop one half into salt solution 1 and the other into solution 2 taken in watch glasses. 2 leave them aside for about 30 minutes. Observe the changes in the stem.

Observation : The piece immersed in a weak solution (solution1) will bend outwards i.e. towards the epidermis and the piece put in strong solution (solution 2) will bend inwards i.e. towards the pith.

Conclusion : In a hypotonic solution (solution 1) the epidermal cells will not absorb water readily due to the presence of cuticle. The pith cells however absorb water (due to osmosis) readily and increase their size. As a result due to the streching of the pith cells the stem bends outwards. In a hypertonic solution (solution 2) the pith cells lose water much faster than the epidermal cells. As a result they (pith cells) shrink resulting in the inward curvature of the stem.

12. Development of turgor pressure due to osmosis

Materials : Collodion sac, 10% sugar solution, water, beaker, thread, stand etc.

Procedure : Prepare a collodion sac as explained in experiment No. 19. Fill the sac with 10% sugar solution, twist the open end bend it back and tie it securely with the help of a thread. Immerse the sac in water contained in the beaker after securing it to a stand. Observe the changing of shape of the sac.

Observation : It is noticed that the collodion sac starts bulging due to the entry of water from the beaker.

Conclusion : The collodion membrane is selectively permeable. It allows water only but not solutes. Since the solvent concentration in the beaker is more, water enters into the collodion sac due to osmosis. As water enters, the sugar solution presses against the inner wall of the collodion sac. This pressure is called the turgor pressure. One to this pressure the collodion sac bulges.

13. Demonstration of endosmosis and exosmosis Using boiled Eggs

Materials : Wide mouth bottles, 10% salt solution, water, boiled eggs

Procedure : Take two boiled eggs and carefully remove the shell. Let the egg be intact without any damage. Select two wide mouth bottles (A and B) where the diametre of the mouth of the bottle should be such that, the deshelled egg can be squeezed through (the mouth) with some difficulty. Let bottle A have distilled water and bottle B have 10% salt solution. Squeeze one egg each into these bottles. Allow them to be immersed in the liquids for 30 minutes. At the end of this period try to remove the eggs from the bottles by inverting these bottles.

Observation : It will be very difficult to take out the egg from bottle A without breaking it as it would have swollen. From bottle B however, the egg comes out easily as it would have shrunken when compared with the original size.

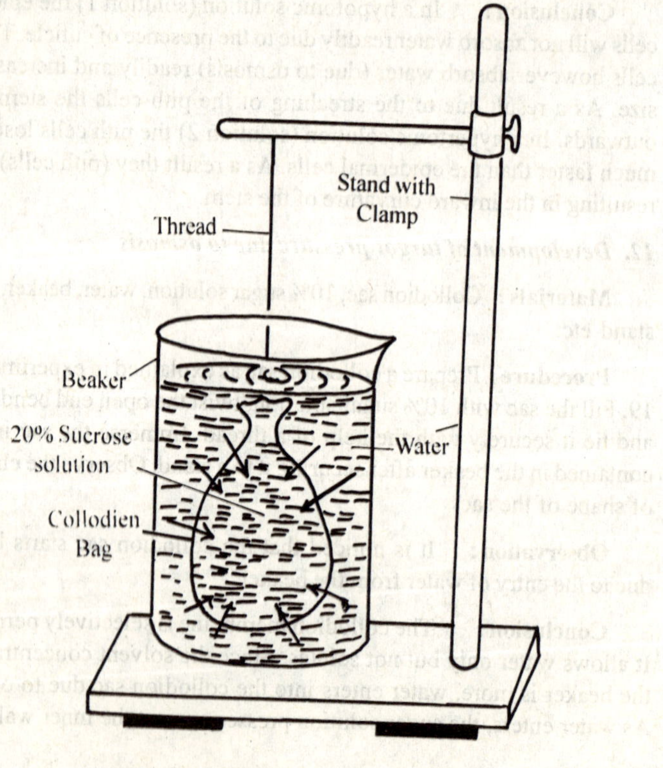

Fig 7.7 Demonstration of Turgor Pressure

Conclusion : The liquid (water) in bottle A is hypotonic when compared with the sap in the egg. The egg is surrounded by a selectively permeable membrane and it acts as an osmotic system and draws water into it. As a result the egg swells. This is called Endosmosis as water enters into a confined system (egg). In bottle B the liquied is a hypertonic solution whose concentration is more than that inside the egg. It is a weak solvent. Hence water goes out of the egg and into the bottle. As a result the egg shrinks in size and can easily come out of the bottle. This is called exosmosis as water comes out of a confined system (egg).

8

WATER RELATIONS

1. Absorption of water by roots

Materials : A potometer, beaker, balsam plant, rubber tube, bent glass tube, measuring cylinder, knife etc.

Procedure : Select a potometer which has a reservoir fitted to a glass jar and a graduated capillary tube attached at another end (see figure). Pull out a healthy balsam plant from the soil with intact roots. Wash the root system under running water to remove all adhering soil particles. Insert the root system into the jar filled with water through a split cork. Now cut off the shoot system a little above the cork and connect the stump to a bent glass tube with the help of a rubber tube. Let the free end of the tube be inserted into a measuring cylinder. Carefully introduce an air bubble into the graduated tube and dip it into a beaker filled with water. Leave the apparatus to stand for some time. Observe the exudation of water into the measuring cylinder and absorption of water through the graduated tube.

Observation : The air bubble moves from one end of the graduated tube to another end as the water is absorbed by the root system. By opening the reservoir the air bubble can be pushed back and the movement of the air bubble can be monitored. The quantity of water absorbed by the plant can be calculated by measuring the distance travelled by the airbubble. The liquied that comes out of the stump gets collected in the measuring cylinder is due to root pressure as a result of water absorption. It may be noted that the amount of water collected in the measuring cylinder is approximatelly equivalent to the amount of water absorbed.

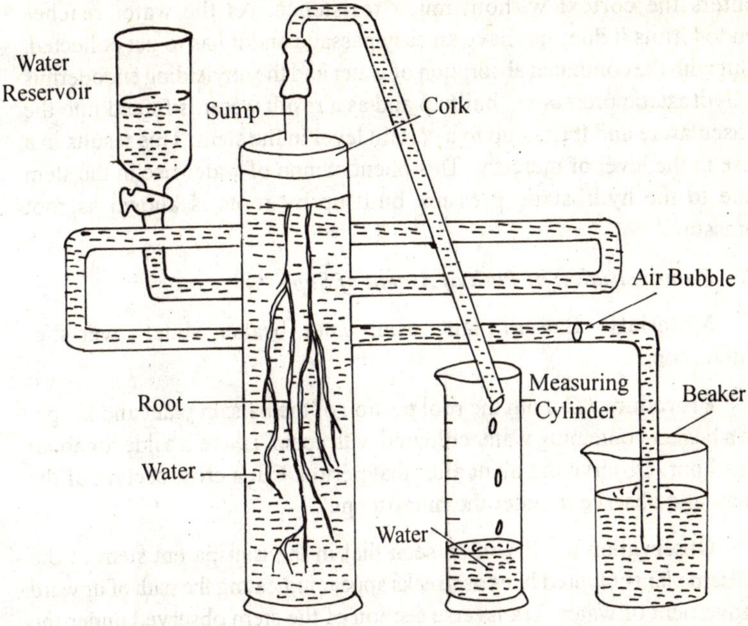

Fig 8.1 Demonstration of Water Obsorption

Conclusion : This experiment demonstrates the absorption of water through roots and its movement into the stem.

ASCENT OF SAP

2. *Root Pressure*

Materials : A healthy potted plant, rubber tube, blade or knife, mercury manometer, water, stand etc.

Procedure : Take a healthy potted plant. Keep it in sunlight. Water the plant well. After some time cut off the shoot a few inches above the soil level and attach a mercury manometer to it with the help of a rubber tube. Fill the tube below the bulb of manometer with water and above the bulb with mercury. Note the initial level of mercury. Keep the apparatus for a while. Note the level of mercury again.

Observation : There is increase in the level of mercury in the manometer.

Conclusion : Water being continuosly absorbed by the roots enters the cortext without much resistence. As the water reaches endodermis it does not have an easy passage and it has to get collected. But with the continued absorption of water and the unyielding endodermis a hydrostatic pressure is built up and as a result water is forced into the vasculature and it rises up to a certain level in the stem. This results in a rise in the level of mercury. This phenomenon of water rise in the stem due to the hydrostatic pressure built up by roots is known as root pressure.

3. *Demonstration of Path of Ascent of Sap*

Materials : Balsam plant eosin, water, beaker, slide, cover slip, microscope.

Procedure : Cut out the root portion of the balsam plant and keep it in a beaker containing water coloured with eosin. Leave it aside for about one hour. Observe the plant after that period. Cut a cross section of the stem and observe it under the microscope.

Observation : It will be seen that in the transparent stem of the balsam plant coloured lines or streaks appear indicating the path of upward movement of water. Transverse section of the stem observed under the microscope shows that only xylem vessels are coloured indicating that eosin coloured water moves up through them.

Conclusion : This experiment shows that ascent of sap in the stem is through the xylem elements

TRANSPIRATION

4. Demonstration of transpiration by bell Jar Method

Materials : A healthy potted plant, glass plate, vaseline oil, cloth or polythene bag, thread etc.

Procedure : Select a healthy potted plant and keep it on a glass plate. Cover the top of the pot with a polythene bag or oil cloth leaving a hole for the plant. Tie the cloth securely to the pot. Cover the pot with a bell jar and apply vaseline to the edges of the bell on the glass plate to make it air tight. Now keep the entire set up in sun light. Observe the apparatus after about 30 minutes.

Observation : Small drops of water appear on the inner side of the bell jar.

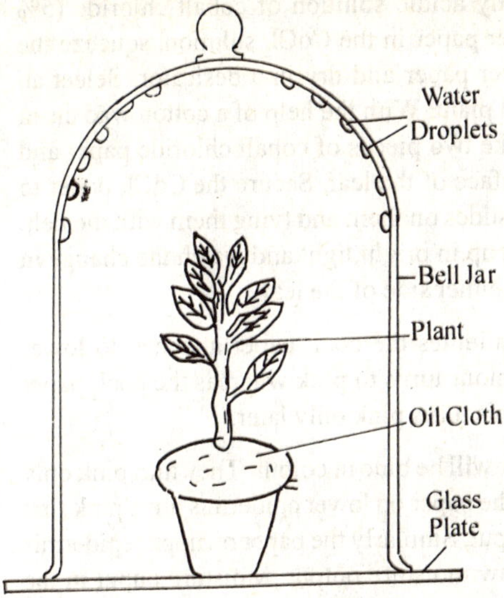

Fig 8.2 Demonstration of Transpiration (Bell jar method)

Conclusion : This experiment demonstrates transpiration. As the water vapouurs come out of the plant they get condensed on the walls of the bell jar and form water drops.

5. *Demonstration of unequal transpiration from the leaf surfaces by Cobalt chloride method*

Materials: A healthy potted plant with dorsiventral leaves, filter paper, cobalt chloride solution (5%) slides, thread, desicator etc.

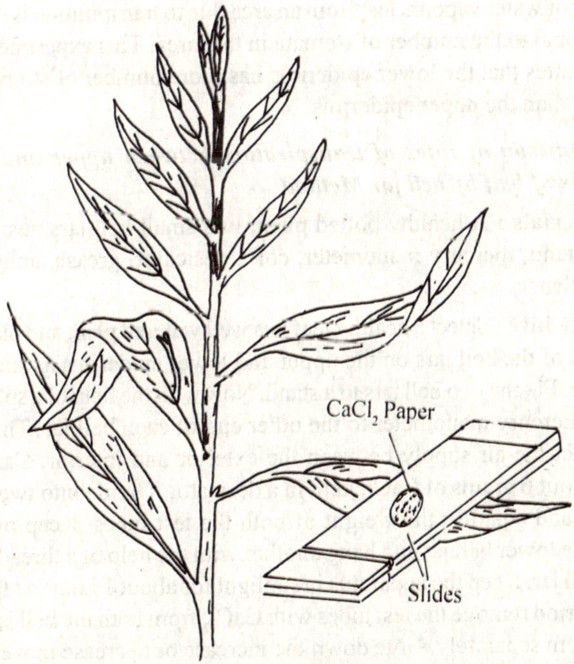

Fig 8.3 Demonstration of Unequal Transpiration (Cobalt Chloride method)

Procedure : Prepare a slightly acidic solution of cobalt chloride (5% solution). Dip pieces of filter paper in the $CoCl_2$ solution, squeeze the excess solution from the filter paper and dry in a desicator. Select an intact leaf in a healthy potted plant. With the help of a cotton wad clean both surfaces of the leaf. Take two pieces of cobalt chloride paper and place one piece on either surface of the leaf. Secure the $CoCl_2$ paper to the leaf by placing dry glass slides on them and tying them with the help of thread. Keep the entire set up in bright light and watch the change in colour of the filter papers on either side of the leaf.

Observation : Within few minutes the $cocl_2$ paper attached to lower epidermis which is blue in colour turns to pink whereas the $cocl_2$ paper attached to the upper epidermis turns pink only later.

Conclusion : Dry $Cocl_2$ paper will be blue in colour. They turn pink only in the presence of moisture. The paper on lower epidermis turns pink first because of more moisture output. Similarly the paper on upper epidermis turns pink later because of low moisture output. Moisture ouput in the form of vapours in due to transpiration. This experiment shows the rate of transpiration to be more in lower epidermis than in upper epidermis. Quantity of water vapours lost from an area due to transpiration is directly proportional to the number of stomata in that area. This experiment also demonstrates that the lower epidermis has more number of stomata per unit area than the upper epidermis.

6. Comparison of rates of transpiration between upper and lower epidermis of leaf by bell jar Method

Materials : A healthy potted plant, two small bell jars, test tubes, thread, stand, mercury manometer, cork, desicator, grease, anhydrous $CaCl_2$, balance

Procedure : Select a healthy leaf in a well watered plant and place the open end of the bell jars on the upper and lower epidermis as shown in the figure. Fix the two bell jars to a stand. Now with the help of a split cork insert a mercury manometer to the other end of each bell jar. This is to disconnect the air supply between the exterior and interior. Carefully weigh about 6 grams of $CaCl_2$ kept in a desicator. Divide into two equal portions and eqaulise the weight of both the test tubes. Keep one test tube in the lower bell jar and hang another with the help of a thread in the upper bell jar. Keep the apparatus in sunlight for about 1 hour. At the end of this period remove the test tubes with $CaCl_2$ from both the bell jars and weigh them separately. Note down the increase or decrease in weight.

Observation : It is noticed that the weight of $CaCl_2$ tube in the lower bell jar has increased more than that of the tube in the upper bell jar.

Conclusion : The rate of transpiration is more in the lower epidermis than in the upper epidermis. As a result $CaCl_2$ kept in the lower bell jar absorbs more moisture and weighs more than the $CaCl_2$ tube kept in the upper bell jar. Higher rate of transpiration in the lower epidermis is due to more number of stomata per unit area than in the upper epidermis.

7. Unequal Transpiration in the lower and Upper epidermis - 4 leaves experiment

Materials : Four fresh dicotyledonous leaves, vaseline, thread, stand etc.

Precedure : Select four healthy dicotyledonous leaves of approximately same age from a single plant. Apply vaseline to the cut end of the petiole and hang them on to a thread streched between two stands by securely tying them. Smear vaseline selectively to the surfaces of the leaf as follows -

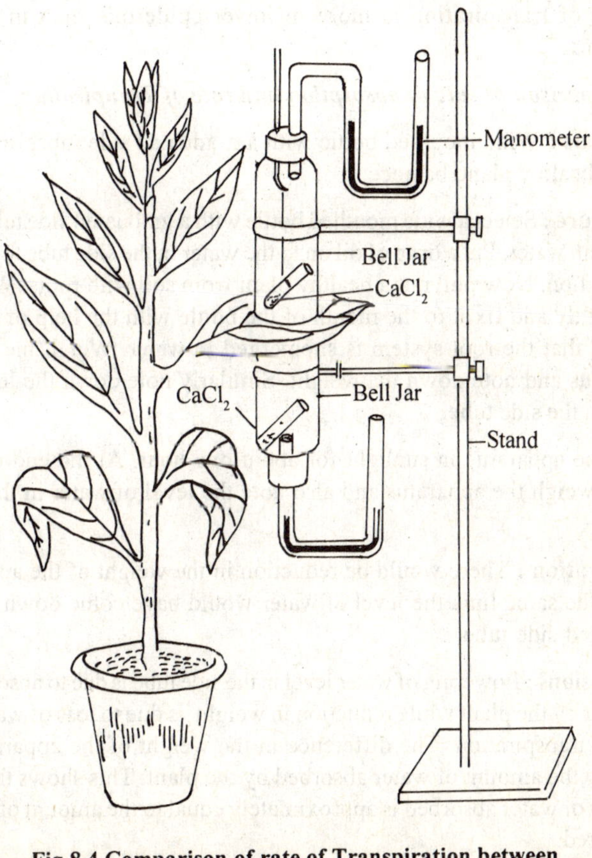

Fig 8.4 Comparison of rate of Transpiration between upper and lower surface of leaf

(a) Leaf 1 : Vaseline on both surfaces
(b) Leaf 2 : Vaseline on only upper surface
(c) Leaf 3 : Vaseline on only lower surface
(d) Leaf 4 : No vaseline (control)

Let the apparatus stand for about 2 hours

Observation : Leaf 4 wilts first followed by leaf 2 and 3, while leaf 1 does not wilt and remains fresh.

Conclusions : Leaf 4 wilts first because of unchecked transpiration from both upper and lower epidermis while leaf wilts next because of unchecked transpiration from lower epidermis. Leaf 3 takes a little longer because transpiration from the lower surface which is always high is checked because of smearing vaseline. Leaf 1 remains fresh for a longer time because transpiration from both upper and lower
epidermis is checked. Since he leaf 2 wilts earlier than leaf 3 it is clear that the rate of transpiration is more in lower epidermis than in upper epidermis.

8. Comparison of rate of absorption and rate of Tranpiration

Materials : Wide mouthed bottle with a graduated side tube, oil, split cork, a healthy plant, balance.

Procedure : Select a wide mouthed bottle with a graduated side tube and fill it with water. Put a drop of oil on to the water in the side tube to check evaporation. Now pull out a healthy plant from soil with roots. Wash it thoroughly and fix it to the mouth of the bottle with the help of a split cork so that the root system is submerged in water. Weigh the entire apparatus and note down the weight. Similarly note down the level of water in the side tube.

Keep the apparatus in sunlight for about one hour. At the end of this period weigh the apparatus and also note the level of water in the side tube.

Observation : There would be reduction in the weight of the aparatus and at the same time the level of water would have come down in the graduated side tube.

Conclusions : Lowering of water level in the side tube is due to absorption of water by the plant while reduction in weight is due to loss of water by way of transpiration. The difference in the weight of the apparatus is equal to the amount of water absorbed by the plant. This shows that the amount of water absorbed is approximately equal to the amount of water transpired.

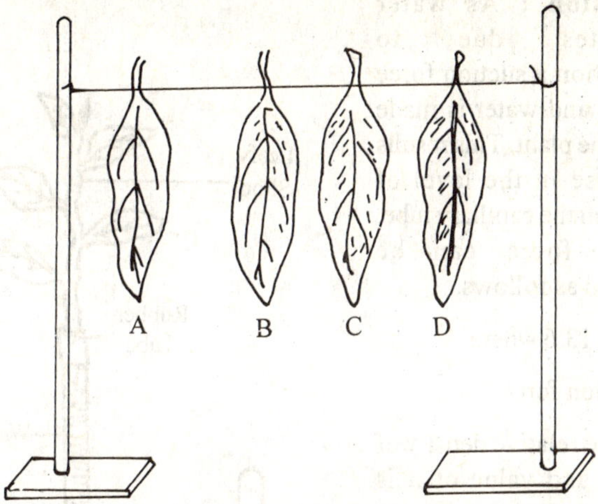

Fig 8.5 Four leaves Experiment

9. Development of suction force due to transpiration

Materials : An open ended glass jar with a side tube, a healthy plant, stand, beaker, mercury, water, glass tube etc.

Procedure : Take a glass jar with a side tube. Fill in water and close all the three ends with the help of corks. To the lower end of the jar fix a capillary glass tube and dip the other end of the tube in a beaker containing mercury. Initially the capillary tube is also filled with water. Fix a plant with roots into the side tube with the help of a split cork. Make the apparatus air tight and keep it in sun light. Observe after about 45 minutes.

Observation : It is observed that the level of mercury rises in the capillary tube.

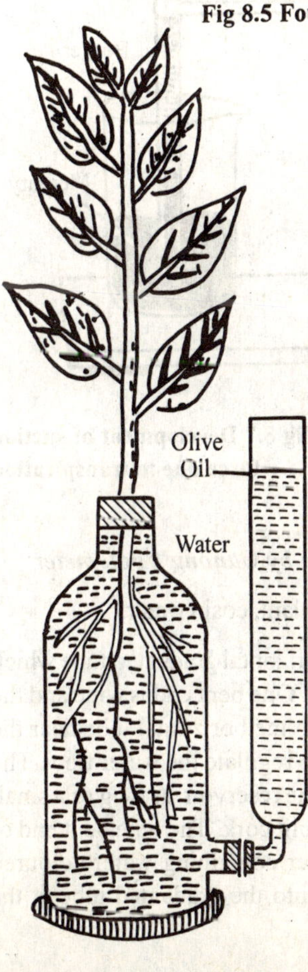

Fig 8.6 Apparatus to compare transpiration and Absorption

Conclusion : As water evaporates due to transpiration a suction force develops and water is made to enter the plant. This results in the rise in the level of mercury in the capillary tube. Suction force can be calculated as follows.

Sf = πr^2 x 13.6 where

sf = suction force

13.6 is the relative density of mercury and value of π is 22/7.

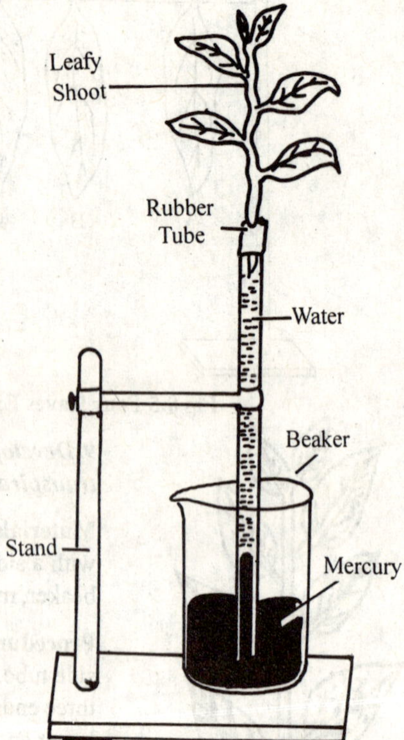

Fig 8.7 Development of suction force due to transpiration

10. Measurement of rate of transpiration by Ganong's potometer

Materials : Ganongs potometer, beaker, plant, eosin water.

Procedure : The apparatus consists of a horizontal graduated tube which is bent in opposite directions at either ends. One bent end is wide and the other end is narrow. A reservoir is fixed to the horizontal tube near the wider end. The reservoir has a stop cock to regulate the water flow. The apparatus is filled with water by opening the reservoir. A twig or a small plant is fixed to the wider arm through a split cork. The othe bent end of the horizontal tube is dipped into a beaker containing water coloured with eosin. An air bubble is introduced into the graduated tube at the narrow end (arm). As transpiration takes place, the air bubble will move from one end of the graduated tube to the other. After it has reached the reservoir, it (reservoir) is opened to push the air bubble back and then its movement can be watched again.

Observation : Record the time that is needed for the air bubble to move from one end to the other; make many observations to obtain the average of time taken to travel the distance.

Conclusion : The Ganong's potometer measures the rate of transpiration indirectly by measuring the rate of absorption. The rate of transpiration is approximately equal to the rate of absorption. By using the following formula the amount of water transpired may be caluclated.

$V = \pi r^2 l$

Where V = value of water, π = 22/7, r = radius of graduated tube, 1 = distance travelled by the air bubble.

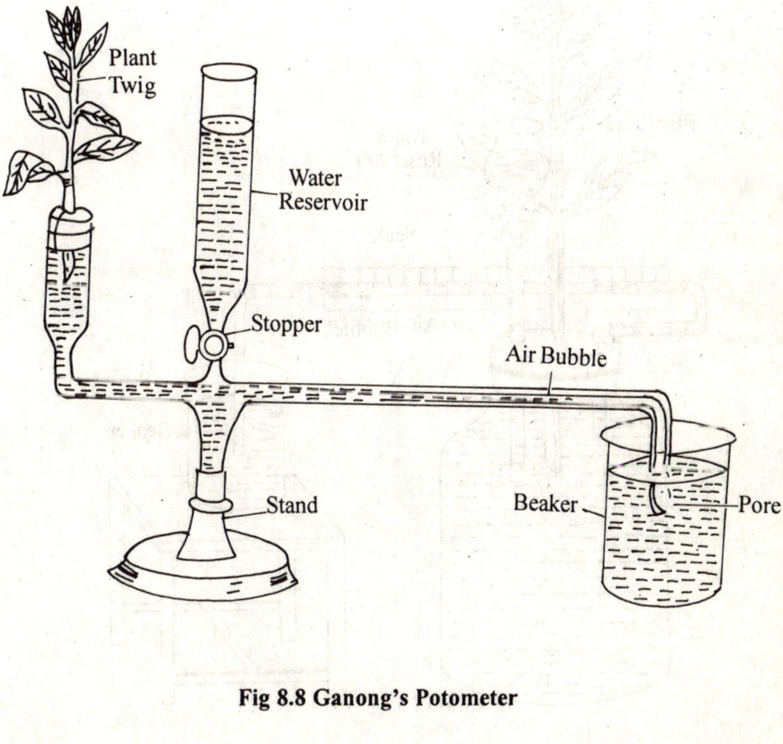

Fig 8.8 Ganong's Potometer

11. Measurement of Transpiration by Farmers' Potometer

Materials : Farmer's potometer, beaker, water, vaseline, plant

Procedure : The Farmer's potometer consists of a wide mouthed bottle. A cork with three holes is fitted on to the mouth of the bottle. One whole is fixed with a thrice bent tube with a scale fixed on to it horizontally, the second hole has a reservoir with a stopper fitted onto it and through the third hole fix a twig under water.

Fill the apparatus with water, open the reservoir so that the tube gets filled with water. Make the apparatus air tight by applying vaseline. Dip the free bent end of the horizontal tube into a beaker containing water. Introduce an air bubble into the tube. Keep the apparatus in sunlight for some time.

Observation and *Conclusion* : Same as in experiment No. 35.

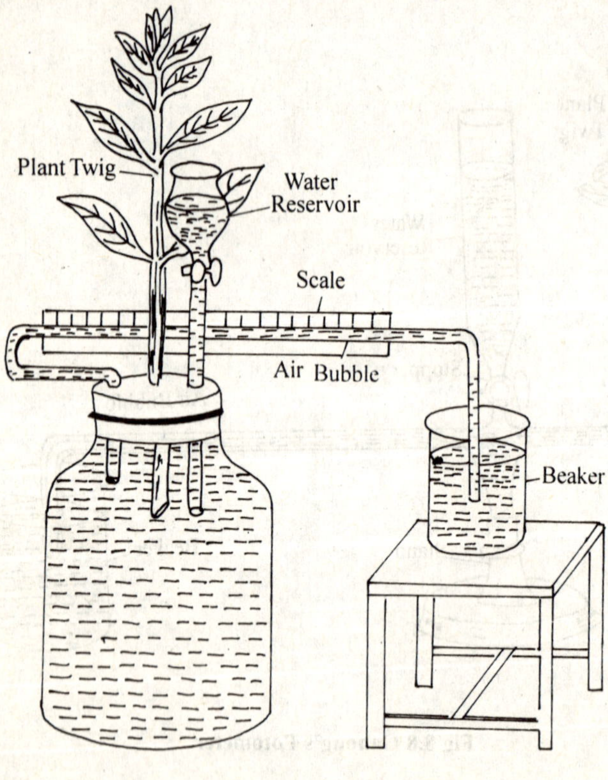

Fig 8.9 Farmer's Patometer

12. Demonstration of Continuity of air spaces in the leaf

Materials : Wide mouthed bottle, cork with two holes, aspirator, bent glass tube, vaseline, a leaf with a long petiole such as *Colocasia*, lotus etc.

Procedure : Take a wide mouthed bottle which is half filled with water. Close the mouth with a two holed cork. Introduce a leaf into the bottle through one hole so that the petiole is dipped in water. Through the other hole fix a bent tube so that the tip of the tube shoudl be few centimetres above the water level. Connect the other end of the tube to an aspirator. Apply vaseline to all connecting points so that the apparatus is air tight. Start the aspirator and observe the cut end of the petiole.

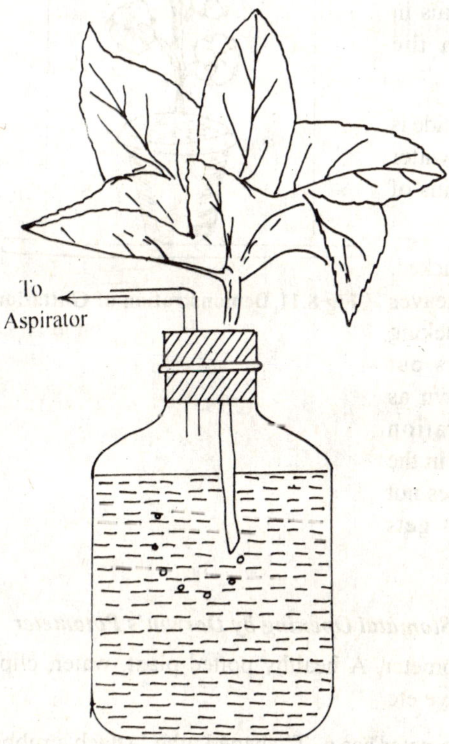

Fig 8.10 Demonstration of Continuity of Air Spaces in Plant

Observations : Air bubbles start coming out of the petiole as long as the aspirator is on.

Conclusion : As the aspirator sucks air the suction is transmitted through the petiols along the intercellular spaces and up to stomata. Stomata suck air and it travels down the petiole and released into water. This is possible only when the aii space is continious within the plant.

13. Demonstration of Guttation

Materials :
Nasturtium plant in a pot, glass plate, bell jar, plastic sheet, vaseline, thread.

Procedure : Keep the potted planted (*Nasturtium* on a glass; plate cover the pot with a plastic sheet with a hole to allow the plant to be exposed. Invert the bell jar on the pot. Let the mouth of the bell jar at the narrow end be covered with a single holed cork. Fix a bent tube through this whole so that one end is in the bell jar while the other end is fixed to an aspirator. Make the apparatus air tight by applying vaseline. Keep the apparatus in sunlight and switch on the aspirator and observe.

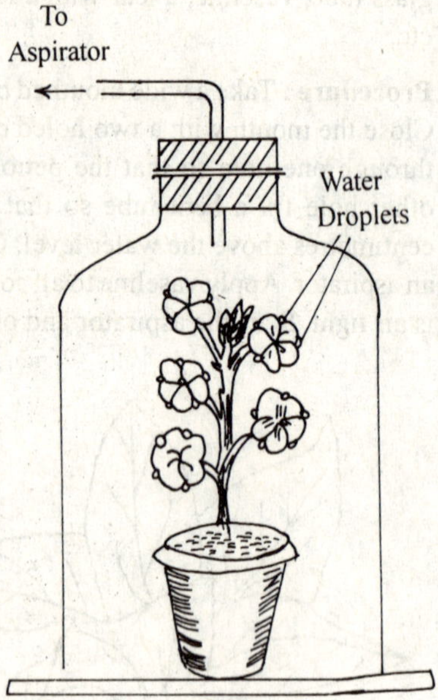

Fig 8.11 Demonstration of Guttation

Observation : As the air inside is sucked by the aspirator water drops appear on the margins of leaves.

Conclusion : As the air is sucked, hydathodes present in the leaves push out water due to the sucking force. Water here comes out physically and it is known as guttation. In transpiration however water comes out in the form of vapours and it does not settle on the leaves, but gets condensed on the bell jar.

14. Measurement of Stomatal Opening by Darwin's Potometer

Materials : Darwin's prometer, A healthy potted plant, water, clips, stopwatch, beaker, adhesive etc.

Procedure : Darwin's prometer has a 'T' shaped tube. Attach a rubber tubing to both the arms of 'T'. Close one of the rubber tubings with the help of a clip while to another attach a small cup. Fix the cup to the lower surface of a leaf of a potted plant with an adhesive. Smear vaseline to the edges of the cup to make it air tight. Dip the stem of 'T' in a beaker containing water open the clip and suck water through it. Water column rises in the stem of 'T' tube. Note the level. Separate the leaf from the plant and after sometime note the level of water in 'T'.

Observation : When the leaf is attached to the plant the level of water in 'T' goes down gradually while the level remains unaffected if the leaf is detached from the plant.

Conclusions : When air is sucked through the rubber tubing its pressure is lowered in the potometer. In the intact leaf stomata are open and transpire, with the result water vapours enter the cup and push down the water column. When the leaf is detached stomata remain closed, transpiration does not occur and water level in the 'T' tube remains same.

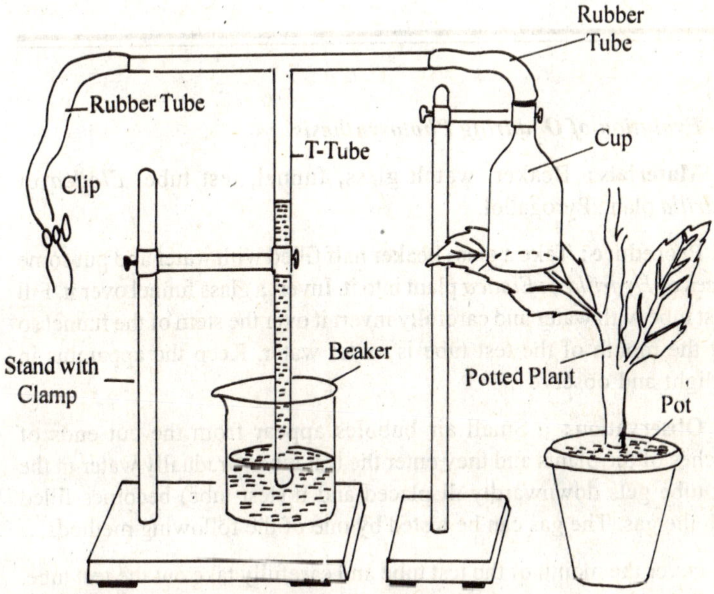

Fig 8.12 Demonstration of Stomatal opening by Darwin's potometer method

9

PHOTOSYNTHESIS

1. *Evolution of O_2 during Photosynthesis*

Materials : Beaker, watch glass, funnel, test tube, *Elodea* or *Hydrilla* plant, Pyrogallol.

Procedure : Take a glass beaker half filled with water and put some pieces of *Hydrilla* or *Elodea* plant into it. Invert a glass funnel over it. Fill a test tube with water and carefully invert it over the stem of the funnel so that the mouth of the test tube is under water. Keep the apparatus in sunlight and observe

Observation : Small air bubbles appear from the cut ends of braches of the plants and they enter the test tube. Gradually water in the test tube gets downwardly displaced and it (test tube) becomes filled with the gas. The gas can be tested by one of the following methods.

(a) cover the mouth of the test tube and carefully take out the test tube. Introduce a burning splinter into the test tube. It glows brightly.

(b) Dissolve some amount of pyrogallol in water. The gas is dissolved in pyrogallol and the level of the water in the test tube rises again.

Conclusion : Both the tests performed above will show that the gass is oxygen. The plants kept in the beaker photosynthesize in the presence of sunlight and release oxygen which comes out of the plant in the form of air bubble.

2. Chlorophyll is necessary for photosynthresis

Materials : Plant with variegated leaves such as *Codeum*, *Coleus*, *Manihot* etc, butter paper, pencil 70% alcohol, potassium iodide solution, water bath

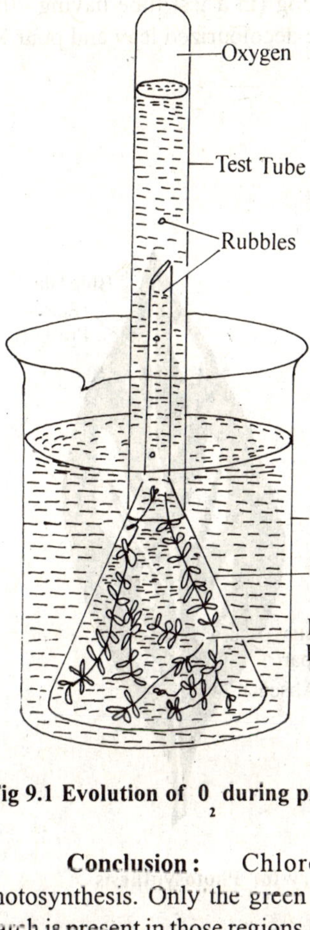

Fig 9.1 Evolution of 0_2 during photosynthesis

Procedure : Select a potted plant with variegated leaves and keep it in darkness for 72 hours so that it gets destarched. Bring out the plant and place it in bright sunlight. Select a leaf with variesated colours and placing a butter paper on the leaf trace the outline of various colours with the help of a pencil. After few hours pull out the leaf from the plant, boil it in 70% alcohol in a water bath so that the leaf gets decolourized. Pour KI solution on the leaf. Observe the change in colour.

Observation : Blue colour appears only in those areas which were green. This can be checked in the outline diagram of the leaf.

Conclusion : Chlorophyll pigment is necessary for photosynthesis. Only the green areas in the leaf photosynthesize and starch is present in those regions. That is why when KI solution is poured they turn blue (indication of starch). Other areas (non green) do not photosynthesize, do not have starch and do not turn blue on contact with KI solution.

3. Light is necessary for Photosynthesis (Ganong's light screen experiment)

Materials : A potted plant, Ganong's light screen, beaker, 7% alcohol, KI solution, test tube, water bath.

Procedure : Keep a potted plant in darkness for about 72 hours so that it is starch free.

Fix a ganong's light screen to one of the leaves and place the pot in bright sunlight. After few hours pluck the leaf from the plant, remove the light screen and decolourize it by boiling (in a test tube having 70% alcohol) it in a water bath. Take out the decolourized leav and pour KI solution.

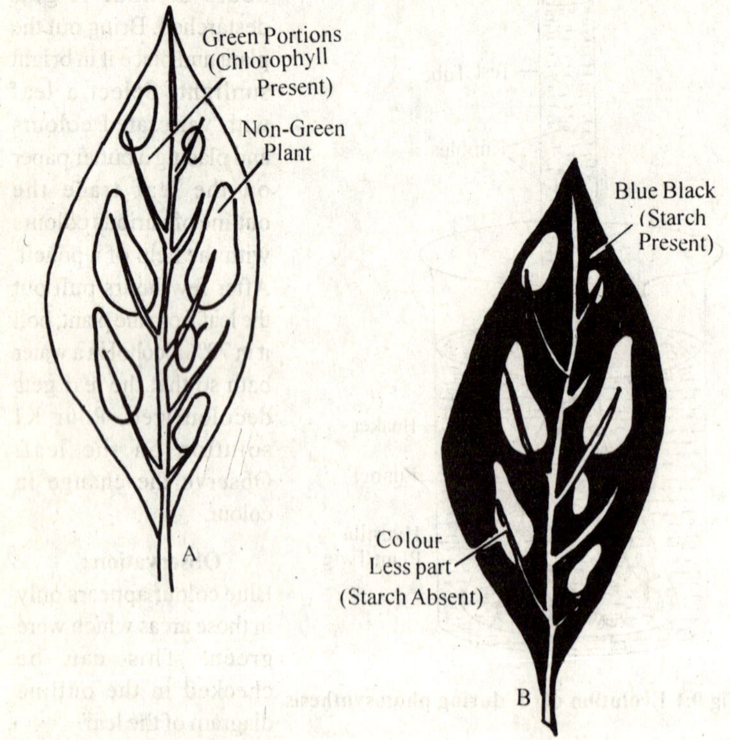

Fig 9.2 Chlorophyll is necessary for Photosynthesis

Observation : The area of leaf covered by light screen remains bleached while the other areas turn blue on contact with KI solution.

Conclusion : The area covered by the light screen does not get light hence starch is not manufactured in that area. As a result it does not turn blue on contact with KI solution. Other areas get light so they manufacture starch and turn blue on contact with KI solution.

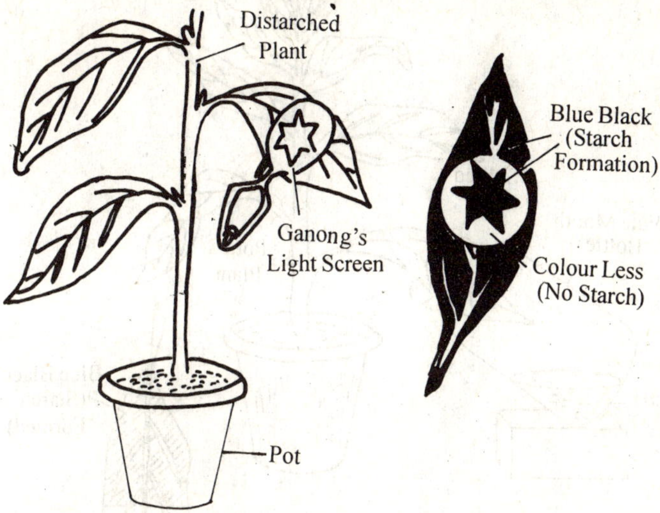

Fig 9.3 Light is necessary for photosynthesis

4. ***CO_2 is necessary for photosynthesis (Mohl's Half leaf experiment)***

Materials : Wide mouthed bottle, split cork, KOH, Healthy plant.

Procedure : Keep a potted plant in darkness for about 72 hours to remove starch. Select a leaf from such a plant and without removing it insert half a portion into a bottle through a split cork. Before inserting the leaf pour some quantity of KoH liquid into the bottle. Make the apparatus air tight by smearing vaseline to the edges of the cork. Keep the apparatus in sunlight for about 2 hours. At the end of the period remove the leaf from the bottle detach it from the plant and boil it is 70% alcohol to decolourize it. Pour Iodine solution and observe the change in colour.

Observation : The part of the leaf inside the bottle will remain bleached whereas the portion outside turns blue.

Conclusion : The portion of leaf inside bottle had all the necessary factors for photosynthesis except CO_2 because it was absorbed by KoH. The portion of leaf outside the bottle turns blue as it has starch manufactured by photosynthesis. It was provided with all the factors including CO_2. This demonstrates clearing that CO_2 is necessary for Photosynthesis.

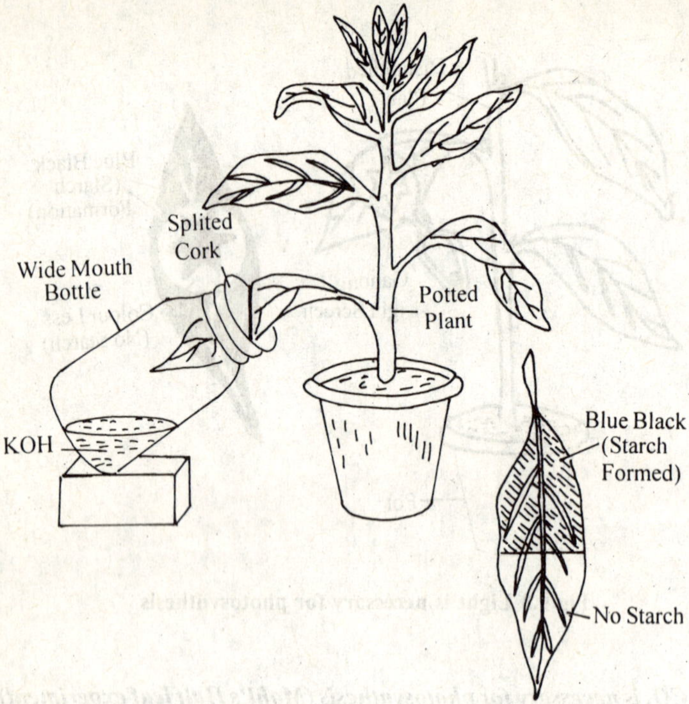

Fig 9.4 Mohl's half leaf experiment

5. *Effect of different wave lengths of light on photosynthesis*

Materials : A healthy potted plant, Ganong's light box, beaker, 70% alcohol, KI solution, test tube, water bath.

Procedure : Ganong's light box is a small rectangular box with removable lid which accomodates glass plates of different colours. Select a leaf from a healthy potted plant which has been destarched (by keeping it in darkness for 72 hours) and insert it into Ganong's light box. Let the box have three colour plates viz. red (660 μm), blue (455 μm) and green (530 μm). Keep the apparatus in sunlight with this arrangement the leaf will be exposed to three wave lengths of light viz red, blue and green of the visible spectrum. After about 2-3 hours remove the leaf from light box detach it from the plant, decolourize it by boiling it in 70% alcohol in a water bath. Pour Iodine solution over the leaf and observe the change in colour.

Observation : Three portions of the leaf were exposed to three different rays of light. The segment of leaf exposed to red light turns deep blue, the part of leaf exposed to blue rays turns light blue and the part of leaf exposed to green light remians almost unchanged

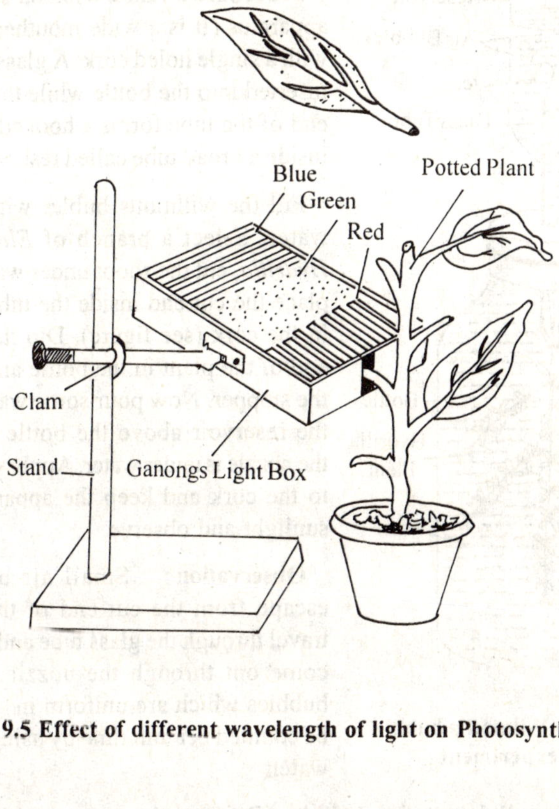

Fig 9.5 Effect of different wavelength of light on Photosynthesis

Conclusion : Part of the leaf exposed to red rays turning deep blue indicates concentration of starch, light blue colour in the region exposed to blue rays is an indication of less starch and the region exposed to green light remaining unchanged indicates lack of starch. This experiment shows that for photosynthesis red and blue rays are necessary while green rays are not needed.

6. Measurement of Rate of Photosynthesis by Wilmott's Bubbler

Materials : Wilmott's bubbler, *Hydrilla* or *Elodea* Plant, pond water, blade, stop watch etc.

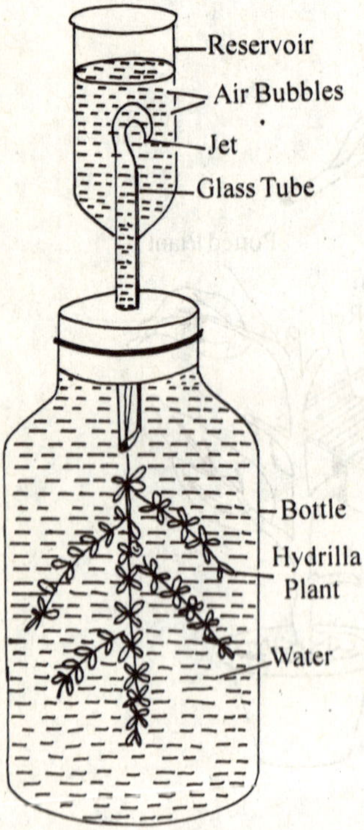

Fig 9.6 Wilmotl's bubble experiment

Procedure : Take a Willmott's bubbler apparatus (It is a wide mouthed bottle with a single holed cork. A glass tube is inserted into the bottle while the upper end of the tube forms a hooked nozzle inside a broad tube called reservoir).

Fill the willmotts bubler with pond water. Select a branch of *Elodea* or *Hydrilla*, cut the shoot under water and place the cut end inside the tube fitted in the cork (see figure). Dip the other end of the plant in the bottle and close the stopper. Now pour some water into the reservoir above the bottle so that the nozzle is under water. Apply vaseline to the cork and keep the apparatus in sunlight and observe.

Observation : Small air bubbles escape from the cut end of the stem travel through the glass tube and finally come out through the nozzle. These bubbles which are uniform in size can be counted per unit time by using a stop watch

Conclusion : Release of air bubbles (O_2) from the cut end of the plant indicates the rate of photosynthesis. The number of bubbles per unit time can be calculated to measure the rate of photsynthesis. This is only an arbitary measurement.

7. Demonstration of presence of starch in chloroplasts

Materials : Filaments of *Spirogyra* or leaves of *Elodea*, Chloral hydrate Iodine solution, slide, cover glass, microscope etc.

Procedure : Mount a filament of *Spirogyra* a green alga under the microscope. Note down the shape of chloroplasts in the cells. Stain the cells with chloral hydrate Iodine solution and observe under the microscope.

Observation : The chloroplasts in *Spirogyra* filament are spiral in shape. After staining the entire spiral band turns blue black. The chloroplasts turn blue black due to the presence of starch.

8. Study of photosynthesis by gass analysis (Ganong's Photosynthometer)

Materials : Photosynthometer, leaves, test tube, caustic potash, carbon dioxide generator.

Procedure : Ganong's photosynthometer consists of three parts - (1) a glass bulb A(2) a graduated tube C and (3) a connecting link B provided with a stop cock. B fits into the neck of A at the lower end and at the upper end the lower end of tube C fits into it. The total volume of the whole apparatus is approximately 103 ml. There are two holes a and b in A and B respectively so that they are at the same level when A fits into B.

Place a few fresh green leaves in the bulb of 'A' together with a little quantity of water. Now invert the graduated tube 'C' close its stop cock and fill it with 5 ml of water. Fix 'C' on to 'B' and close the stoplock of latter and fill water into lower half portion of 'B' and carefully covering it by hand dip into a beaker with water. Clamp the tube so that water in the beaker is on level with hole b. Open the stop cock of B as well as of 'C' and connect 'C' to a carbon dioxide generator. Stop the Co_2 entry into tube 'C' as soon as the water in tube 'C' falls to the level of water in beaker. Tube 'C' now has 5 ml of Co_2 (by displacing 5 ml of water). Close both the stop cocks ('C' and 'B\) and fit the whole thing ('B' and 'C') on to the bulb 'A'. Now the openings a and b (in 'A' and 'B' parts) are on the same level. this will equalise the pressure inside the bulb A to the atmospheric level. Twist 'B' slightly now so that a and b move apart and the connection becomes airtight. Place the apparatus in sunlight and open the stopcock of 'B' to allow CO_2 into bulb A (with leaves). Allow the apparatus to stand for 6 - 8 hours. At the end of this period, remove 'B' and 'C' together from 'A' ande dip them in water. Separate 'B' from 'C' underwater and open the stop cock of 'C' so that water enters into it up to zero mark. Fill a test tube with caustic potash and connect it to 'C' near the stop lock with the help of a rubber tube provided with a pinch cock. Lift the graduated tube ('C') from water by closing the lower end with hand and open the pinch cock as well as the stop cock to allow potash fluid to flow into 'C'. Shake thoroughly and drain the potash back to the tube and close the pinch cock as well as the stop cock. Now, still closing the lower end of 'C' with hand dip it into water, disconnect the test tube with caustic potash. Place the tube (in water) in such a way that the zero mark is on level with water. Take out the hand and water will rise into the tube up to a mark.

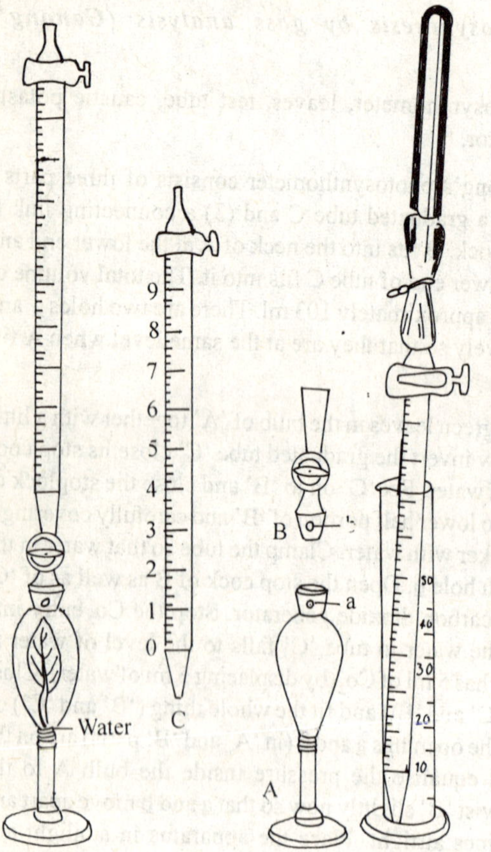

Fig 9.7 Ganong's Photosynthometer

Observation and Conclusion : The amount of water which rises up in the tube is equal to the amount of CO_2 used up for photosynthesis. Originally 5 ml of CO_2 is present. Now if water rises up to 2 ml, it means 2 ml of CO_2 is used up for photosynthesis and 3 ml is left unused. (It is possible that this may not give an accurate amount of CO_2 used in photosynthesis, because CO_2 is slightly soluble in water).

9. Extraction of Chlorophyll Pigments By Solvent wash method

Materials : Fresh leaves (spinach, tecoma etc), or 3 grams of dry leaf powder, petroleum ether, Acetone, methyl alcohol,methanolic KOH, Separating funnel, stand, beaker, pestle, mortar etc.

Procedure : Take 5 gms of fresh leaves of spinach or *Tecoma.* Wash them thoroughly. Wipe off moisture with the help of a filter paper. Place them in a pestle and mortar and grind with 50 ml of ice cold acetone.

Collect the liquid and filter. Discard the residue and collect the filterate. The filterate (acetone extract) consists of chlorophyll a, chlorphyll b, *carotene* and *xanthophyll.*

Instead of fresh leaves, leaf powder may be used. Collect fresh leaves and dry them in an oven. Powder the dry leaves. Weigh 3 gms and soak it in 40 ml of 80% acetone containing a pinch of CaCo3. Stir the solution and filter. The filterate acetone extract consists of all the pigaments mentioned above.

Add 60 ml of petroleum ether to 40 ml of acetone extract in a separateing funnel. Gently rotate the funnel. Slowly add 75 ml of distilled water along the sides of the funnel to avoid emulsification. Rotate the funnel again gently and fix it to a stand. Gradually two layers of the liquid are formed. Of these the lower is acetone - water layer. This may be discarded. The upper layer (petroleum ether layer) consists of all the pigments. Add 50 ml of distilled water gently to the petroleum ether layer. Again two layers are formed. Discard the lower layer. The upper layer may be washed again with 50 ml of distilled water and discarding the lower layer.

Add to the petroleum ether layer 50 ml of 92 % methyl alcohol. Gently rotate the separating funnel. Allow it to stand. Tow layers are formed viz. the lower methyl alcohol layer and the upper petroleum ether layer. Draw the two layers into separate beakers. Lower layer consists of chlorophyll b and xanthophyll. The upper layer consists of chlorophyll a and carotene. Further procedure for two layers is as follows -

1. **Upper Petroleum ether layer :** Take 30 ml of this layer and to it add 15 ml of freshly prepared 30% methyl alcoholic potassium hydroxide solution in a separating funnel. Rotate gently. Two layers are formed. Add 30 ml water and shake. The upper petroleum ether layer consists of carotene while the lower methanolic KoH layer consists of chlorophyll a. Collect these two layers in separate tubes.

2. **Lower Methyl Alcohol Layer :** Take 50 ml of methyl alcohol layer in a separating funnel and add 50 ml of diethyl ehter. Mix them thoroughly by rotating the funnel. Two layers are formed. Discard the lower methyl alcohol layer. The upper layer has thepigments. Add 15 ml of methanolic KOH to the upper layer. Shake gently for about 10 minutes. Add 30 ml of distilled water and shake again. Two layers are formed. The lower methanolic KoH fraction consists of chlorophyll b, while the upper diethyl either layer consists of xanthophyll. Collect them separately.

Observation and Conclusion : The four pigments separated can be characterised by checking their peak absorption spectra. The following is a flow chart for the process given above.

Leaves of Spinach, Tecoma etc
(5 gms)
↓ Grind in 50 ml of ice cold acetone and filter
Leaf extract
↓ Add 60 ml of petroleum ether and rotate gently
Separating funnel
↓
Wash gently with 75 ml of distilled water
↓
Two layers of the liquid
↓

Upper petroleum ether layer | Lower acetone - water layer discard

Upper petroleum ether layer:
↓
Wash with distilled water three times and discard lower layer every time
↓
Add 50 ml of 92% methyl alcohol
↓ Roate gently
Two layers
↓

Upper petroleumether layer | Lower methyl alcohol layer. Collect in alcohol funnel

Upper petroleumether layer:
↓
Add 15 ml of methanolic KoH and rotate gently
↓
Shake for 10 minutes ,
↓
Add 30 ml of distilled water and shake
↓
Two layers
↓
Petroleum ether layer (Carotene) | Methanolic KOH layer (Chlorophyll $\underline{a}$)

Lower methyl alcohol layer. Collect in alcohol funnel:
↓
Add 50 ml of diethyl ehter
↓
↓
Wash with 30 ml of water
↓
Two layers
↓
Upper ether layer ↓ | Lower methyl alcohol layer Discard

Add 15 ml of methanolic KoH
(30% and Shake

↓

Add 30 ml water and shake

↓

Two layers

↓

Upper diethyll ether layer (**xanthophyll**)	Lower Methanolic Koh Layer (**Chlorophyll b**)

10. Separation of leaf pigments by circular paper Chromotography

Materials : Two large petridishes, Whatman No. 1 filter paper, acetone, chlorophyll extract.

Procedure : Take a circular piece of whatman No. 1 filter paper and apply the chlorophyll extract to the centre of the paper in the form of a spot. Be careful not to allow the spot to spread. Use a hair dryer if necessary to dry the spot. Cut the filter paper from the periphery to the centre in the form of a small strip to form a 'wick'. Now place the filter paper over a small petridish (diameter must be smaller than that of the filter paper) containing about 30 ml of 20% acetone. Bend the wick in such a way that it dips in the acetone. Cover a large petridsh over the paper to prevent evaporation. Keep it for about 30 minutes. Watch the progress of the solvent. Remove the filter paper before the solvent reaches the outer periphery (of the paper). Dry it and observev the rings that appear. Measure the distance of the rings from the centre.

Observation and Conclusion : Four rings of coloured streaks appear. Of these the one closest to the centre of the filter paper is chlorophyll b, followed by chlorophyll a, xanthophyll and carotene. Rf values for each ring may be calculated as follows.

$$Rf = \frac{\text{Distance travelled by ring of pigment}}{\text{Distance travelled by the solvent}}$$

Rf value will be always less than 1 as distance travelled by the solvent is more than the distance travelled by the compound.

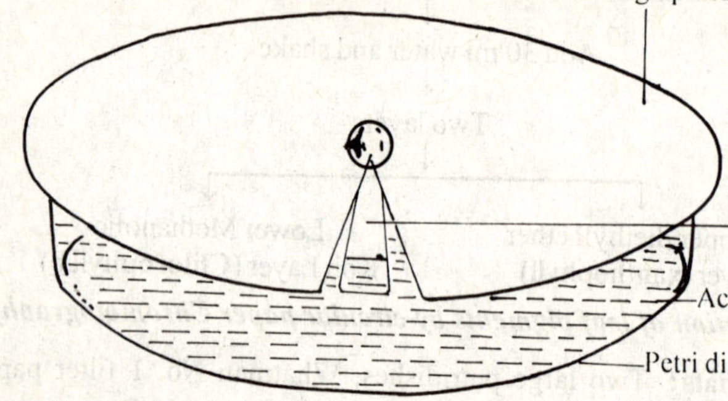

Fig 9.8 Circular paper chromatography for separation of leaf pigments

11. Separation of leaf pigments by Ascescending paper Chromatography

Materials : Cylindrical or/rectangular jar with a tight fitting lid, chlorophyll extract, whatman No.1. Filter paper, petroleum ether, acetone, hook or thread, micro pipette, pencil etc.

Procedure : Prepare the chlorophyll extract as explained in experiment 48. Cut out a long strip of whatman No. 1 filter paper and draw across a pencil line about 2-3 centimeters from the end. With the help of a micropipetle draw a little quantity of concentrated chlorophyll extract and spot it on the centre of the pencil line. Allow the spot to dry (use a hair dryer if necessary). Apply some more quantity of the extract to the same region. Be careful not to allow the spot to expand on the paper.

Now select a jar of suitable size (longer than the filter paper strip) with a tight fitting split cork. Pour 50 ml of the solvent (petroleum ether : acetone -= 100 : 12) into the jar. Fix the upper end of the filter pape (i.e away from the spot region) to the split cork and insert the paper into the jar so that the spot end dips into the solvent below. The paper should be dipped in such a way that the spot is above the level of the solvent. Apply vaseline to the edges of the cork. Allow it to settle for about 30 - 45 minutes. Keep observing.

Observation : The solvent migrates up along the paper strip and carries the spot along with it. Gradually 4 different coloured spots apper at different distances from the original spot. Remove the paper from the jar before the solvent reaches the upper edge of the pape and dry it.

Conclusion : The four spots that appear are (starting from a distance closest to the original spot) . Chlorphyll b, chlorophyll a, xathophyll and carotene. Calculate the Rf values as given in experiment 49.

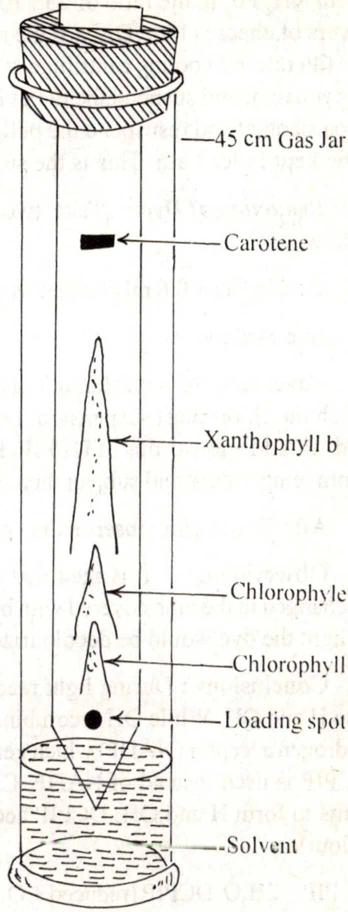

Fig 9.9 Ascending paper chromatography for separation of leaf pigments

12. Demonstration of Hill's reaction in isolated Chloroplasts (Dye reduction method)

Materials : Spinach leaves, Sodium phosphate buffer (Nacl solution 0.25 ml, 0.1m potassium dihydrogen phosphate), 0.1%, 2-6 dichlorophenol indophenol (DCPIP), Icc bath, water bath, pestle and mortar, cheese cloth, test tubes, aluminium foil etc.

Procedure : There are two major steps (1) preparation of chloroplast suspension and (2) reduction of the dye DCPIP.

(1) Preparation of Chloroplast Suspension : Prepare an ice bath and cool all the reagents, chemicals and glass ware used in the experiment. Weigh about 10 grams of fresh spinach leaves, wash them thoroughly and cut them into pieces. Remove, petiole, midrib and other hard materials and collect only the lamina portion. Macerate this tissue in pestle and mortar kept in an ice bucket. Alternatively the tissue can be homogenised in a waring blender with the operations conducted in a cold room.

Add 20 ml of sodium phosphate buffer at pH 6.5 (mix 0.24 m Nacl and 0.1m KH_2 Po_4 in the ratio of 25 : 10 ml). Filter the suspension with two layers of cheese cloth. Discard the residue and collect the filterate. Spin the filterate in a cooling centrifuge at 3000 rpm for 15 minutes. Discard the super natant and suspend the pellet in buffer and spin again. Discard the supernantant and resuspend the pellet in buffer. Stir it and transfer it to a tube kept in ice bath. This is the suspension of isolated chlorplasts.

(2) Reduction of Dye : Take two test tubes and add ingredients as follows :

A. 5.0 ml buffer + 0.6 ml distilled water + 2.0 ml of chloroplast suspension

B. Same as above.

Cover tube 'A' with aluminium foil or black paper so that no light can reach the chloroplast suspension. Let tube B be uncovered. To both 'A' and 'B' add a few drops of DCPIP. Keep both the tubes in water bath at room temperature and subject them to strong illumination.

After some time observe the contents of both tubes.

Observation : It is seen that the blue colour of the dye remains unchanged in the tube covered with black paper while in the tube exposed to light the dye would be decolourızed.

Conclusions : During light reaction of photosynthesis water splits into H and OH. While OH recombines to form water, H is accepted by a hydrogen acceptor (NADP) which gets reduced. In the present experiment DCPIP is used instead of NADP. Chloroplasts receive light and water splits to form H and OH. DCPIP accepts H, gets reduced and becomes colourless

DCPIP + $2H_2O$ DCPIP (reduced + O_2

This experiment demonstrates that isolated chloroplasts can carry out photosynthesis even in a cell free medium.

10

RESPIRATION

1. Demonstration of Anaerobic respiration

Materials : Petridish, test tube, germinating seeds, mercury, KOH crystals, forceps etc.

Procedure : Fill a test tube with mercury and invert it over a petridish also containing mercury. Now lift the test tube lightly and introduce few germinating seeds (gram, pea etc with testa removed) with the help of a forceps. The seeds will rise to the top of the test tube. Keep the test tube back in its position and allow the apparatus to stand for about 45 minutes.

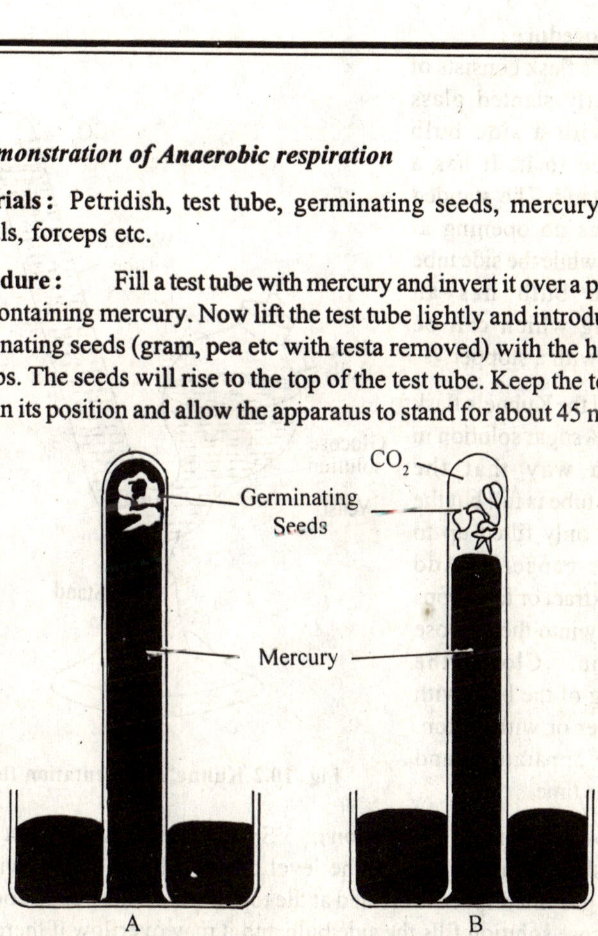

Fig 10.1 Demonstration of Anaerobic Respiration

Observation : A gas starts building up at the test tube and it pushes the mercury down. Now introduce a piece of caustic potash. This absorbs the gas and the level of mercury goes up and fills the test tube.

Conclusion : Germinating seeds have a high rate of respiration. As they are kept in mercury they do not get oxygen and hence respire anaerobically. Since introduction of KOH absorbs the released gas it (gas) must be CO_2. This experiment demonstrates that CO_2 is released during anaerobic respiration

2. *Demonstration of Fermentation (Kuhne's fermentation flask)*

Materials : Kuhne's flask, yeast extract or toddy, 5% glucose solution, KOH pellets

Procedure :
Kuhne's flask consists of a slightly slanted glass tube with a side bulb attached to it. It has a basal stand. The standed tube has no opening at the top while the side tube with a bulb has an opening which can be closed with a stopper.

Fill the Kuhne's flask with 5% sugar solution in such a way that the slanted tube is full but the bulb is only filled up to half its capacity. Add yeast extract or few drops of toddy into the glucose solution. Close the opening of the bulb with a stopper or with cotton. Let the apparatus stand for some time.

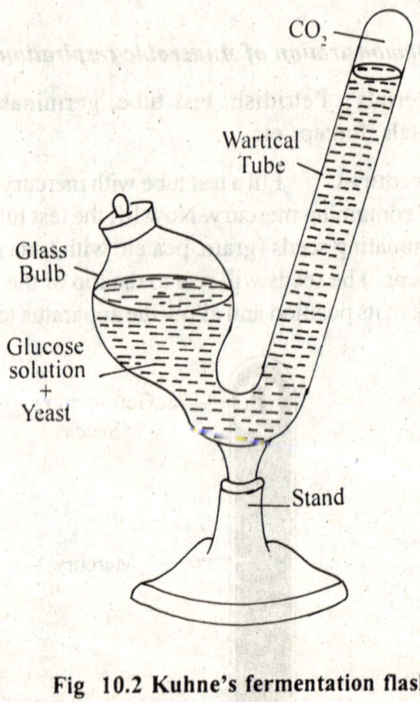

Fig 10.2 Kuhne's fermentation flask

Observation and Conclusion : Small bubbles of gas start rising in the slanted tube and push the level of the solution below with the result a gas filled space is created at the top. If this is allowed to continue the glucose solution fills the side bulb and it may overflow if there is no stopper. Introduce a piece of KOH into the tube through the bulb so that it reaches the gas filled top, the gas is absorbed and the glucose solution once again moves to the top of the tube. The gas must be CO_2 as it is absorbed by KOH. This shows that CO_2 is released during fermentation. This is different from anaerobic respiration because here the respiratory substrate is outside the cell (glucose solution). Alcohol is formed during fermentation.

3. *Release of Co_2 during aerobic respiration (conical flask method)*

Materials : Conical flask, A twice bent tube, test tube, thread, germinating seeds, KOH solution, vesline beaker, saline water.

Procedure : Take a conical flask and place some germinating seeds in it. Close the mouth with a single holed cork. Introduce a twice bent tube through the hole. Take a small quantity of KOH solution in a test tube and hang it inside the flask with the help of a thread. The thread will be held tight with the help of the cork. Let the free end of the bent tube be dipped in a beaker containing saline water. Apply vaseline to the cork to make it air tight. Let the apparatus stand for some time

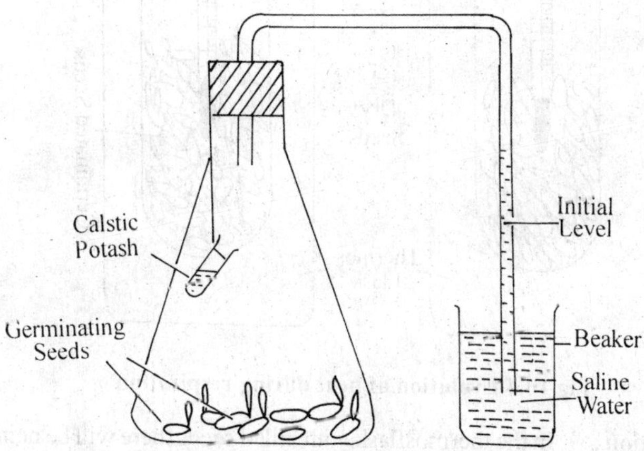

Fig 10.3 Release of CO_2 during aerobic respiration

Observation : Water level in the tube rises after some time.

Conclusion : As the germinating seeds respire to release Co_2 which is absorbed by KoH. This creates a kind of suction pressure which is transfered along the tube into the beaker containing saline water making it to rise.

4. *Evolution of heat during respiration*

Materials : Two thermos flasks, germinating seeds, mercuric chloride, rubber cork etc.

Procedure : Take two thermos flasks. Keep germinating seeds in one and mercuric chloride treated (killed) seeds in the other. With the help of single holed corks introduce a thermometer into each of these flasks and close the mouth (of the termo flask). Note the inital temperature in both the flasks. After 45 minutes note the temperature again. Record your observations.

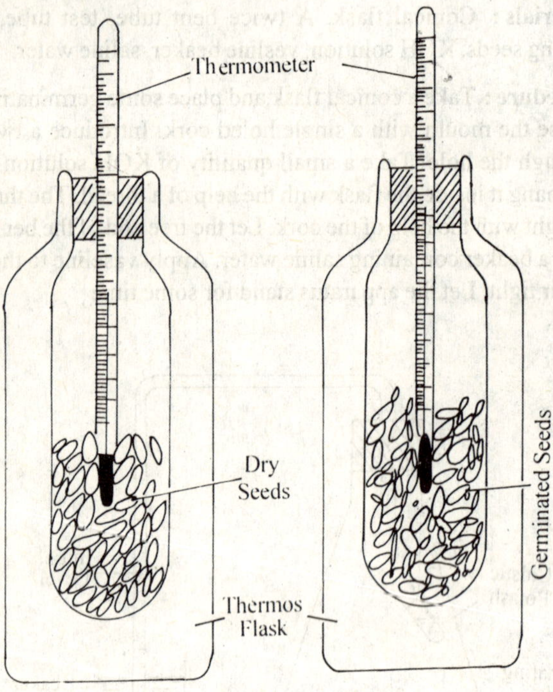

Fig 10.4 Evolution of heat during respiration

Observation : In the thermosflask with killed seeds there will be neither increase or decrease in termperature as recorded by the thermometer. In the other flask having germinating seeds there will be an increase in temperature as recorded by the thermometer.

Conclusion : During respiration there is release of energy, while most of it used for the synthesis of ATP, some amount escapes as heat and this results in increased temperature in the flask with germinating seeds. In the other flask where the seeds are killed, there is no increase in temp as there is no respiration.

5. Demonstration of use of O_2 during Respiration

Materials : Conical flask, beaker, bent glass tube, trough, mercury, germinating seeds, caustic potash stand etc.

Procedure : Keep some germinating seeds in a conical flask. Also keep a small vial containing KOH inside the flask. Now close the mouth of the conical flask with the help of a cork having a single hole. Through this hole introduce a twice bent glass tube. Let the the free end of the tube be inserted in a beaker containing mercuryh. Allow the experiment to stand for some time.

Observation : Mercury kept in the beaker rises in the glass tube up to 15 cm height and then stops.

Conclusion : Germinating seeds respire and release Co_2 which is immediately absorbed by KOH. This creates a suction force and mercury rises up in the glass tube. After reaching a height of 15 cms rise of level of mercury stops indicating the stoppage of respiration and consequently the production of CO_2.15 cms height of mercury in the tube is approximately one fifth of normal height at which atmospheric pressure maintains it. This means that the seeds must have utilized 1/5th of air. The share of O_2 in atmospheric gases is also 1/5th. It may therefore be assumed that seeds have absorbed O_2 (for respiration) which accounts for 1/5th of air.

The rise in level of mercury is equivalent to the amount of CO_2 produced. This means the production of Co_2 during respiration is about 1/5th of the volume of atmosphere. During respiration the amount of O_2 utilized is equivalent to the amount of Co_2 released ($C_6H_{12}O_6 + 6O_2 + 6H_2o$ $6Co_2 + 2H_2o$). Hence this indicates that 1/5th of volume of air i.e, the entire O_2 content is used up. As there is no further amount of O_2 left in the conical flask respiration stops and the rise in the level of mercury also stops.

6. Demonstration of evolution of CO_2 during aerobic respiration by gas flow method

Materials : Potted healthy plant, bell jary black cloth, wide mouth bottles, 'U' tube, soda lime, bent tubes, lime water, baryta water, aspirator, vaseline etc.

Procedure : Take a bell jar and invert it over a healthy potted plant. Close the top of the bell jar with a two holed cork. Connect two bent tubes to these holes. To one tube connect a bottle containing baryta water. This bottle also should be fitted with a two holed cork. Insert a bent tube into this bottle with the other end of the tube being connected to an aspirator. Now the other bent tube from the bell jar should be connected serially to two bottles having lime water and then onto a 'U' tube containing soda lime. Smear vaseline to all the connections and cover the bell jar with a black cloth.

Set up the apparatus in such a way that air necessary for respiration first enters the 'U' tube followed by two bottles containing lime water and then into bell jar. As a result the air that enters the bell jar will be free of CO_2 as it gets absorbed by lime water in bottles. As the bell jar is covered by black cloth there will be no photosynthesis and hence no chance of CO_2 released during respiration to be utilized for photosynthesis. The air now enters the bottle containing baxyta water and then to the aspiration.

Observation and conclusion : Baryta water turns milky showing that the air from bell jar (with the plant) has CO_2. The air entering the bell jar is Co_2 free while the air leaving bell jar has CO_2. Hence the CO_2 must have been released by the plant during aerobic respiration.

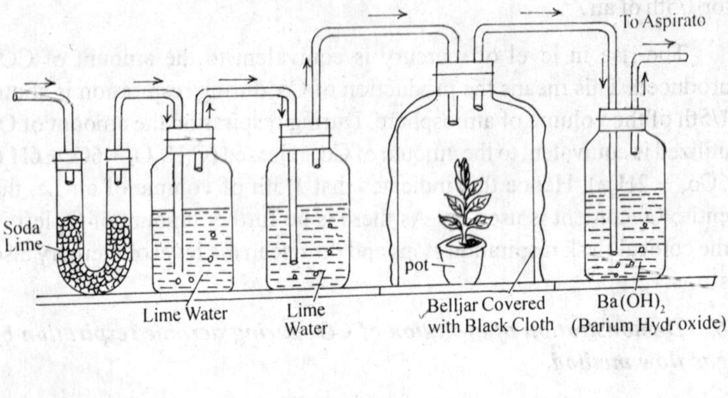

Fig 10.5 Evolution of CO_2 during aerobic respiration (Gas flor method)

7. ***Evolution of CO_2 during an aerobic respiration***

Materials : Same as experiment 57 and pyrogallol.

Procedure : Set up the experiment in the same manner as in 57, except instead of two bottles having lime water, replace lime water, in one bottle with pyrogallol. As the air enters thsese bottles, one (lime water) absorbs CO_2 and the other (pyrogallol) absorbs O_2 so that the air entering bell jar has neither CO_2 nor O_2. Still the air that leaves bell jar has CO_2 indicating respiration in the absence of O_2. This demonstrates anaerobic respiration.

8. · ***Measurement of Repiratory quotient (RQ) of different respiratory substances using Ganong's respirometer***

Materials : Ganong's respirometer. Water, saline water, stand, cotton, different germinating seeds such as wheat, rice (carbohydrate), castor, mustard (fat), pea (protein), succulent plant such as *Opuntia* or *Bryophyllum.*

Procedure : Ganong's respirometer basically consits of a vertical graduated tube connected a stopper. The neck of the bulb as well as the stopper have a small hole each so that they are in line at a particular level. The other end of the graduated tube is attached to a levelling tube with the help of rubber hole. Take some quantity of saline solution and pour it into the levelling tube so that both tubes get filled about one half their capacity. Now open the stopper and keep some wet cotton in the bulb together with some germinating seeds (any one mentioned above - in materials). Place the stopper and rotate it slightly in the mouth of the bulb so that the holes in them are in line. The air pressure inside the bulb now equals that of outside. Adjust the levelling tube (lowering or raising) in such a way that the level of saline water in the graduated tube reaches zero level Rotate the stopper now to cut off contact with external air. Let the apparatus stand for some time.

Observation : The Level of saline solution in the graduated tube remains stationary

Conclusion : No change in the level of saline solution indicates that the amount of O_2 used is equivalent to the amount of CO_2 released. There will be no change in volume of air in the bulb. Hence it does not press down nor suck the saline solution from the graduated tube. Thus the RQ is equivalent to 1 or unity. RQ is calcualted as follows.

$$RQ = \frac{\text{Amount of } Co_2 \text{ evolved}}{\text{Amount of } O_2 \text{ used}}$$

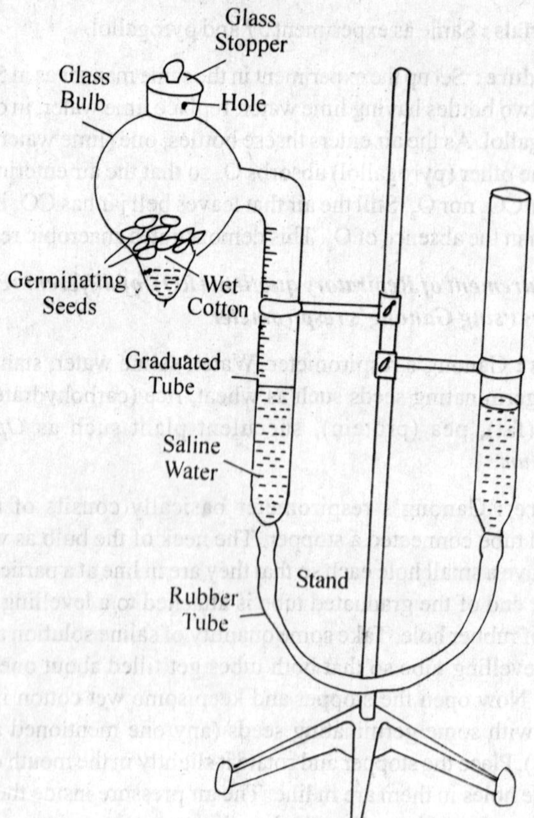

Fig 10.6 Ganong's respirometer

In case of **Carbohydrates** (Wheat, Rice,) such as sugars vol of CO_2 evolved is equivalent to CO_2 used.

$$C_6HO_6 + 6O_2 + 6H_2O \rightarrow 6CO_2 + 12H_2O$$

$$RQ = \frac{\text{Amount of } CO_2 \text{ evolved}}{\text{Amount of } O_2 \text{ used}} = \frac{6}{6} = 1 \text{ or unity}$$

When **fats** are respiratory substrates (castor, mustard etc) the level of saline water in the graduated tube will rise. Fats have less O_2 and hence require more O_2 for oxidation. O_2 used in respiration is more than CO_2 evolved and RQ will be less than unity. When palmition is the respiratory substrate -

$$2C_{51}H_{98}O_6 + 145O_2 \longrightarrow 102CO_2 + 98H_2O$$

$$RQ = \frac{102}{145} = 0.7$$

When the respiratory substrate is a protein (pea, gram etc) the level of saline solution will rise because proteins also require more O_2 in respiration than the amount of CO_2 evolved. As a result volume of air in the bulb gets decreased and the saline water rises being sucked by the air. (Usually RQ for proteins will be 0.8 - 0.9)

In case of **organic acids** being respiratory substrates, more amount of Cos will be released than O2 utilized. As the volume of air increases in the bulb it presses down the saline solution and its level will go down. Here RQ will be more than unity.

$$C_4H_6O_5 + 3O_2 \longrightarrow 4CO_2 + 3H_2O$$

$$RQ = 4/3 = 1.33$$

Caluclation of actual RQ in an experiment using Ganong's respirometer

1. Note the initial level of saline solution in the graduaated tube (l_1)
2. Final level (l_2)
3. The difference between the two $l_1 - l_2 = V_1$ (The value of V_1 may be negative or positive)

Insert a piece of KOH into the graduated tube. As the total CO_2 is absorbed the level will rise further. This rise may be called l_3.

The difference between l_2 and l_3 $(l_3 - l_2) = V_2$

V_2 denotes the actual volume of CO_2 evovled during respiration. While $V_2 + V_1$ gives the volume of CO_2 used during respiration.

9. Comparison of Respiration and Photosynthesis

Materials : Two long necked flasks, petridishes, stands, mercury, KOH pellets, green leaves, petals etc.

Procedure : Take two flasks A and B. Place few fresh moist green leaves in flask 'A' and moist petals in 'B'. Invert the flasks over troughs of mercury so that the neck is immersed in mercury. Fix the flasks to separate stands with the help of clamps. Keep the apparatus under diffuse light for about 10 - 12 hours. At the end of this period introduce KOH pellets into both the flasks and observe.

Observation : There is very little rise in the level of mercury in flask 'A' while in flask 'B' there is a considerable rise. In flask 'A' green leaves photosynthesize and utilize CO_2 rleased during respiration. Hence there is little if any free CO_2 left in the flask. In flask 'B' where petals are kept the entire CO_2 is in the flask when KoH is introduced it obsorbs CO_2 and hence there is rise in the level of mercury.

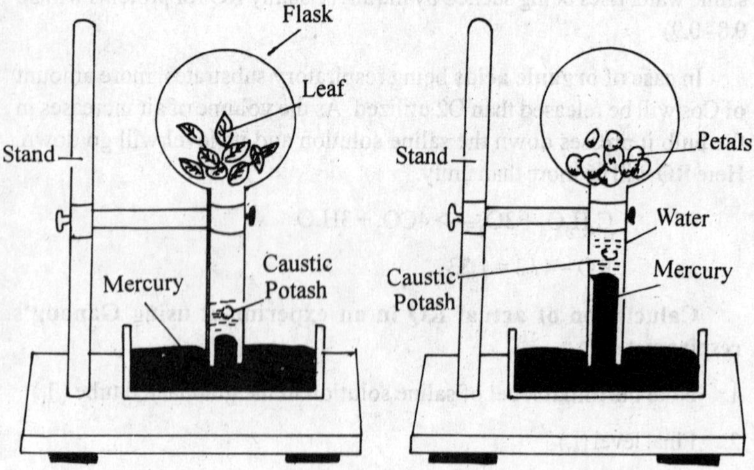

Fig 10.7 Comparison of Respiration and Photosynthesis

11

ENZYMES

1. *Test of Oxidase*

Materials : Fresh apple, tray,

Procedure : Cut thin slices of apple and place them on a tray. The pieces turn brown after some time. This shows that the phenolic compounds get oxidised in the presence of atmospheric oxygen due to the action of enzyme oxidase.

The enzyme oxidase catalyses the oxidation of the substrate by removing hydrogen from them and transfering it to molecular oxygen as follows.

$$AH_2 + O \quad A + H_2o$$

2. *Test for Peroxidase*

Materials : Potato tuber, Buchner's funnel, beaker, mortar, Buchner's funnel, muslin cloth, test tubes, 1% guaiacum (Benzidine) solution, $3H_2o_2$.

Procedure : Crush a fresh potato in a mortar. Filter the slurry through two layers of muslin cloth in a Buchner's funnel. Tkae the filterate (3 ml) in a test tube and add 0.5 ml of 1% Benzidine (dissolve 2 gms of guaiacum in 100 ml of absolute alcohol). Immediately add to the test tube a few drops of 3% H_2O_2 solution. Note down the change in colour of the mixture.

Observation : The colour rapidly changes to blue

Conclusion : Phenolic compounds get oxidized rapidly in the presence of H_2O_2. This shows the presence of peroxidase in the potato tubers. If boiled potato (the enzyme will be denatured) is taken blue colour will not appear. The reaction is as follows :

$$AH_2 + H_2o_2 \; A + 2H_2o$$

3. *Test for amylase*

Materials : Fresh potato, starch solution, germinating barley grains, citrate - phosphate buffer (pH 5.6), Iodine solution, test tube, spirit lamp, water etc.

Procedure : This involves two steps (i) preparation of enzyme extract and (ii) assay for enzyme activity.

Preparation of enzyme extract : Mash germinating barlye grains in cold citrate phosphate buffer (pH 5.6) and filter the extract through two layers of chees cloth. The filterate serves as the enzyme source. Keep it in ice bath.

Enzyme assay : Take two clean test tubes A and B. Add 2ml of enzyme extract to each of the test tubes. Boil the enzyme extract in A in a water bath. Cool it to room temperature. Now to bath tubes add freshly prepared potato starch solution (1 ml) and shake gently. Keep the tubes aside for 10 minutes. Test the contents of A and B separately at periodic intervals with Iodine solution for the presence of starch.

Observation and Conclusion : In tube A the contents always turn blue indicating the presence of starch. In tube 'B' the blue colour will be intense in the beginning but later it turns lighter and finally no colour formation will be there. This indicates the gradual digestion of starch due to the activity of amylase. In tube 'A' the enzyme will not be active as it gets denatured due to boiling.

12

NITROGEN METABOLISM

1. Demonstration of nitrates in green plants :

Materials : A healthy potted plant, slide, cover slip calcium nitrate solution (1.5%), diphenylamine (1%) , microscope.

Procedure : Select a healthy potted plant. Add to the pot 100 ml of 1.5 % calcium nitrate solution. Keep the plant in darkness for 72 hours and then transfer it to bright light. Cut transverse section of lamina and petiole of the leaf and mount them in few drops of diphenylamine sulphate solution (1% diphenylamine dissolved in 50% H_2SO_4). Cover them with cover glass and observe under the microscope.

Observation : A deep blue colour appears

Conclusion : The colour indicates the presence of nitrates.

2. Experiment to demonstrate that atmospheric N_2 is not directly absorbed by green plants :

Materials : Glass troughts (one big and one small), bell jar, dilute sulpuric acid, petridish, funnel tube, sterilized N_2 free sand distilled water.

Description of the apparatus : The apparatus consists of a large glass trough (A) which is half filled with dilute sulphuric acid. A bell jar (B) is inverted inside the trough and placed on three corks so that it is raised above the floor of the trough. A small trough (C) is placed inside the bell jar on a large petridish so that it is not in contact with H_2SO_4 in the large trough. Trough 'C' is filled with water and a small glass pot is kept inside the trough (C). The inverted bell jar has a small inlet tube in the side which is connected on the inside to trough 'C' and on the outside to a funnel. This facilitates addition of water from time to time. The top of the bell jar is closed with a cork having an inlet tube which allows CO_2 inside the bell jar.

Procedure : Collect some legume seeds and estimate their total N_2 content by chemical analysis collect the ash also and estimate the N_2 content. Take some quantity of sterilized N_2 free sand and add it to the small glass pot inside trough 'C' together with some quantity of ash. Sow some legume seeds (the seeds from the same plant on which chemical analysis has been done) in the pot and allow them to germinate. Let them grow into small plants. After a definite period of time (15-12 days) pull out the plant along with roots. Estimate the total N_2 present in the whole plant. Compare it with the amount of N_2 already estimated in the seeds.

Observation and Conclusion : Comparison of the N_2 content shows that the plant does not have more N_2 than the seeds. This shows that the plant has not absorbed gaseous N_2 from atmosphere. Higher plants can only absorb N_2 in the form of nitrates or nitrates. In this experiment the sand from which roots absorb nutrition is N_2 free. The only available form of N_2 for plants is gaseous N_2. Dilute H_2SO_4 absorbs traces of ammonia if any in the bell jar.

13

GROWTH AND HORMONES

1. Measurement of Growth by arc auxanometer

Materials : Arc auxanometer, weight, thread, a potted plant.

The apparatus : The arc auxanometer consists of a tripod stand and a pulley connected to a pointer which moves on an arc scale. A thread is made to pass over the pulley tightly. One end of the pulley is tied to a weight.

Procedure : Take a healthy potted plant and tie its tip to thread which is held over the pulley in such a way that the thread does not slacken. Keep it aside and observe.

Observation and Conclusion : As the stemp top grows, it will push the thread up and the tight thread will turn the pulley. The pulley in turn will move the pointer over the graduated arc scale. The extent of movement from the initial to the final in a given period of time can be measured. The distance moved by the pointer does not give the actual linear growth, because the arc is divided into 90° (one quarter of a circle i.e., 360 / 0.25) and not linear measurements. Further the measurement on the graduated arc is magnified because of the length of the pointer. The actual growth is calculated as follows-

Inital position of pointer = A

Final position of pointer = B

Actual distance travelled by the pointer during a given period of time = B - A

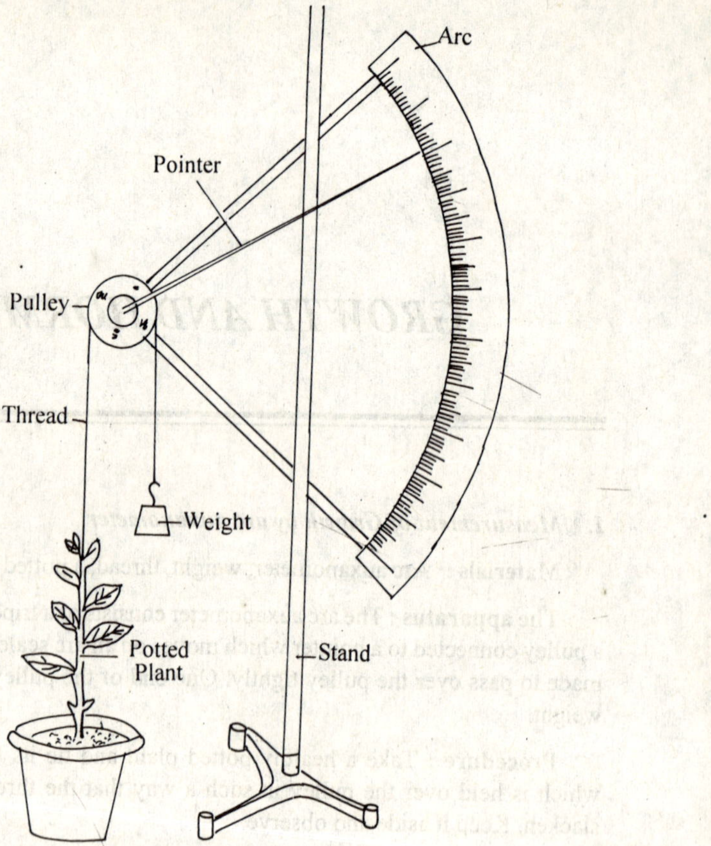

Fig 13.1 Arc Auxamometer

Acutal growth = B - A x Radius of pulley

in length length of pointer (from the centre of the pulley

For example if the radius of pulley is 10 cms and the length of the pointer is 50 cms the magnification is 10 times. If the pointer has moved 5 cms in 20 hours the actual growth is as follows.

5 x 10 = 1 cm in 20 hours

50 Therefore Growth per hour = 0.5 mm

2. *Measurement of growth by Pfeffer's Auxanometer :*

Materials : Pfeffer's auxanometer, potted plant, thread etc.

The apparatus : Pfeffer's auxanometer consists of a vertical stand and a self rotating drum over which a smoked paper is glued. The drum rotates on its axis through a clock work mechanism at the rate of one turn per every hour. Attached to the stand on a shaft are two pulleys one small and the other bitg. The smaller pulley is attached to the central axis of the larger pulley. A thread passes over the small pulley, one end of which is tied to a weight. Another thread also passes over the large pulley with weights tied to both ends. A pointer is attached to one end of this thread slightly above the weight. This pointer touches the smoked paper wrapped around the rotating drum.

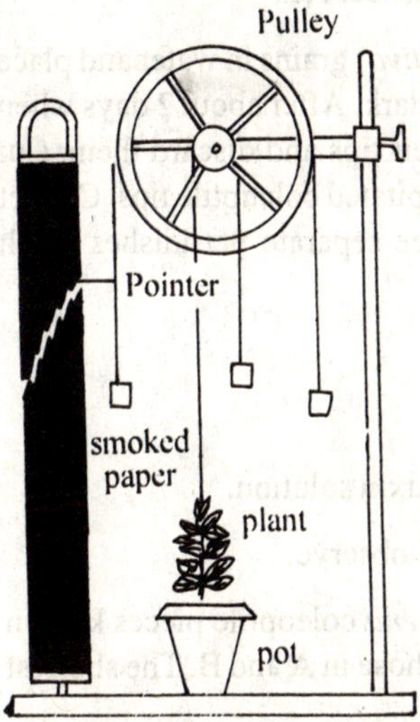

Fig 13.2 Pfeffer's Auxanometer

Procedure : Take a healthy potted plant and tie its tip to the free end of the thread coming over the small pulley. Observe.

Observation and Conclusion : As the stem tip grows the small pulley moves together with large pulley. This results in the pointer continuously marking a spiral line on the rotating drum's smoked paper.

The perpendicular distance between two adjoining spirals is the hourly growth of the plant. Actual growth rate may be calculated taking into consideration the radii of both the pulleys and the rate of rotation of drum.

3. *Measurement of growth by space markers*

Materials : Indeliable ink (India ink), germinating seeds

Procedure : Select few germinating seeds where radicle has emerged out into the tap root. Using a thin nib mark graduations at regular intervals a little behind the root tip. Observe.

Observation and Conclusion : It is observed that the distance between the graduations have increased. This increased distance is the actual linear growth of the root.

4. *Effect of Auxins (Indole acetic acid - IAA) on plant growth*

Materials : Oat grains, Auxin solution (10 mg/l of IAA), 10% sucrose solution, water, circular filter paper, graph sheet etc.

Procedure : Soak some oat (*Avena sativa*) grains in water and place them on a wet filter paper in a petridish in dark. After about 2 days when coleoptiles are 2 - 3 cms long, remove their tips and discard them. Cut sections of about 5 mm long from the decapitated coleuptile tips. Collect about 15 sections, and keep them in three separate petridishes (each having five sections) as follows :

Petridish A Water

Petridish B Sucrose solution

Petridish C Sucrose solution + 1 ml auxin solution.

Keep the petridishes back in darkness and observe.

Observation : It is observed that *Avena* coleoptile pieces kept in petridish C have grown much longer than those in A and B. The shortest length is found in A.

Conclusion : Coleoptile elongation is minimum in 'A' (water) as it has no nutrition. While slight growth is found in B because of sucrose solution. In 'C' however these is maximum elongation as auxins promote coleoptile growth.

Note : Different concentrations of IAA may be tried to find out their effect on growth

5. *Effect of Gibberellins on plant Growth*

Materials : Seeds of pea, tomato, lettuce etc, petridish, circular filter paper, gibberellic acid solution (100 mg/litre), measuring cylinder etc.

Procedure : Germinate seeds in dark on wet filter paper placed in a petridish. When the seeds have germinated for about 48 hours, transfer them to petridishes - (a) filter paper with water and (b) filter paper with gibberellic acid solution. Keep the petridishes in light for 72 hours and observe.

Observation : It is seen that hypocotyls in the petridish with gibberellic acid solution become much more elongated than the one with only water.

Conclusion : Gibberellin has an effect of elongation of cell and hence the hypocotyls become longer in length.

6. Effect of Cytokinin on expansion of Cotyledons

Materials : Seeds of cucumber, pumpkin or ash gourd, petridish, filter paper, 10 mg kinetin solution, water, graph paper etc.

Procedure : Germinate the seeds on wet filter paper placed on a petridish in dark. Select germlings with about 2 - 3 mm of hypocotyl and remove the seed coat. Separate the cotyledons and float them on petridishes - (a) containing only water and (b) containing kinetin solution. The abaxial surface of the cotyledons should be in contact with kinetin solution. Keep them aside for 72 hours and observe.

Observation : It is seen that cotyledons kept in kinetin solution have expanded much more than those kept in water. This can be tested by placing them on graph paper.

Conclusion : Kinetin a growth hormone is known to induce rapid cell division thus bringing about expansion of the cotyledons.

14

PLANT MOVEMENTS

1. *Geotropism (Horizontal Klinostat)*

Materials : Klinostat, a healthy plant

The appratus : Horizontal klinostat essentially consists of a clockwork mechanism. Instead of the hands at the region of the dial there is an axis to which is attached a small metallic tumbler. The clock is fitted to a stand. When the clock is working the tumbler slowly rotates. It remains stationary when the clock is not working.

Procedure : A small potted plant is kept in the tumbler as shown in the figure. Observe what happens to the plant under the following conditions

(a) when the tumbler is rotating (clock working)
(b) When the thumbler is not rotating (clock not working)

Observation : When the tumbler is stationary i.e., clock not working, after few days the shoot bends up while the root bends down (downward curvature of the root may be observed by removing the soil). When the tumbler is rotating, the plant continues to grow horizontally i.e, parallel to ground. There is no upward curvature in the stem and downward curvature in the root.

Conclusion : When the clock is not working auxins accumulate in the lower surface (ground facing side) of stem and root. Auxin accumulation has opposite effects on stem and root. In stem auxin accumulation increases cell division while in root it decreases cell division. As a result of this, shoot bends up because there is more growth on the lower surface than the surface away from ground. In root the auxin accumulated lower surface has less growth than the surface away from gravity hence it bends down. The unequal accumulation is due to gravity.

When the tumbler is rotating there will not be gravitational effecti on any part of the plant as auxins can not accumulate in any one surface. As the plant is also rotating there is equal distribution of auxins. Hence the plant (shoot or root) does not bend but grows horizontally.

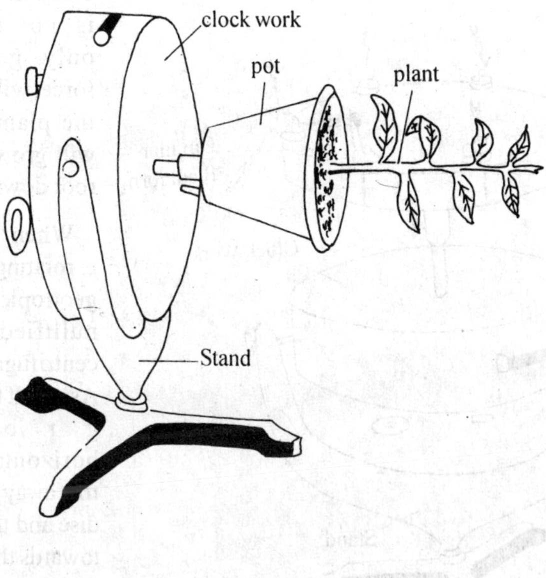

Fig 14.1 Horizontae Klinostat

2. *Effect of Geotropism (Vertical Klinostat)*

Materials : Vertical klinostat, germinated seeds with hypocotyl emerging, cork, pin, etc.

The apparatus : A vertical klinostat is essentially like a horizontal klinostat except that the face of the clock work mechanism will be directed up. In addition, a disk is attached to the rotating axis instead of tumbler. When the clock is operating the disc rotates. Corks are fitted at regular intervals near the periphery of the disc.

Procedure : Attach a few germinated seeds to the cork with the help of pins. Observe the behaviour of seedlings under the following conditions :-

(a) When the disc is rotating slowly
(b) When the disc is not rotating
(c) When the disc is rotating fast

Observation : When the disc is rotating slowly i.e 2.5 turns per hour the roots will diverge and shoots converge at an angle of 45°. The bending is due to the fact that centrifugal force (rotating force) and gravitational force act together and bring about the slanting growth.

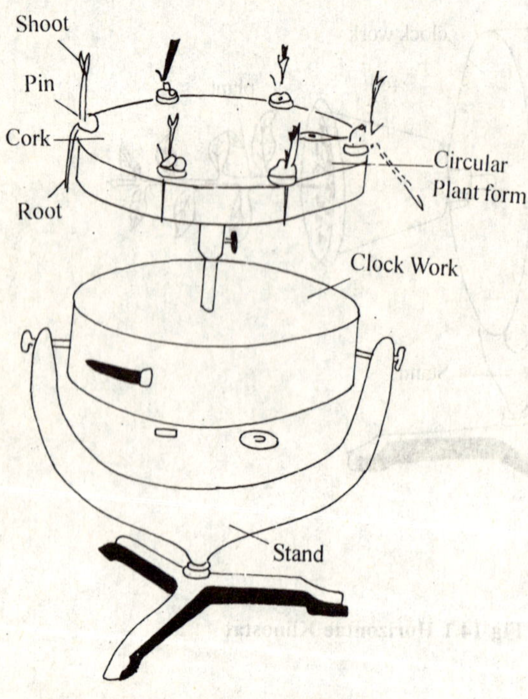

Fig 14.2 Vertical Klinostat

When the disc is not rotating only geotropic force will act on the plant. Shoot will grow up and root down.

When the disc is rotating fast the geotropic force is nullified by the centrifugal force. As result the plant g r o w s horizontally the root away from the disc and the shoot towards the centre of the disc.

3. Phototropism (Heliotropism)

Mateials : Phototrophic chamber (box), A potted plant.

A phototrophic chamber is a small wooden box painted black inside. It has a hinged door at the top (to open and close) and a small window on the side.

Procedure : Keep a potted plant in the phototrophic chamber close the lid. Keep the box in such a way that the window of the box is directed towards the source of light. Let the box stay in place for about 2 - 3 days.

Observation : It is observed that the shoot bends towards the window of the box, comes out and then once again bends up and grows vertically.

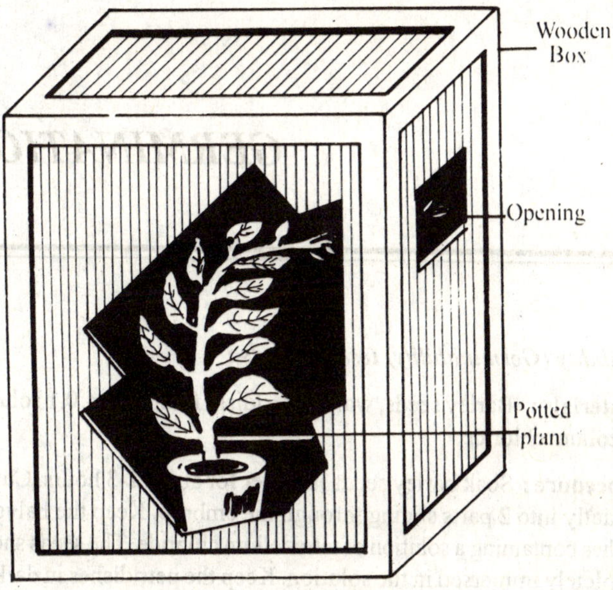

Fig 14.3 Phototropic Chamber

Conclusion : The plant gets light only through the side window. Unidirectional light falling on the side of the stem destroys the auxins on that part of the stem. Auxins accumulate in the shaded region. As a result there is more growth in the shaded region than the lighted region and the shoot bends and comes out of the window of the box. When the shoot comes out of the box the light strikes the upper surface of the horizontal stem making auxins to concentrate on the lower surface. Hence the stem bends up then onwards as the light is not undirectional on the apex it grows up. This is known as phototropism i.e., growth towards light. While shoots are positively phototropic roots are negatively phototropic i.e. they grow away from light (towards gravity)

15

GERMINATION

1. *Viability (Germinability) test for seeds :*

Materials : Barley seeds, water, petridish, filter paper, 0.1 solution of tetrazolium chloride.

Procedure : Soak barley seeds in water for about 2-3 hours. Cut the seed exactly into 2 parts slicing through the embryo. Keep the halves in petridishes containing a solution of tetrazolium chloride. The seeds should be completely immersed in the solution. Keep the petridishes in dark for about 45 minutes and observe.

Observation : Some of the seeds have stained embryos while others are colourless.

Conclusion : Viable seeds respire and can change the colourless tetrazolium compound to a coloured (reduced) compound formazon. Formazon is red in colour and thus the embryos are stained. In non viable seeds there is no respiration and hence the tetrazolium salt remains colourless. Percentage of germination of seeds can be calculated as follows :-

Number of seeds used for experiment = A

Number of seeds stained = B

Percentage germination B/A x 100

Microchemical Tests

16

CARBOHYDRATES

1. *Test for Cellulose*

Materials: Paper or Cotton fibres, Iodine solution, water 75% sulphuric acid, slides, cover slips, microscope etc.

Procedure: Place a few cotton fibres or very thin shreds of paper in a drop of water. Observe the cell wall under the microscope. Add a few drops of Iodine and a few drops of H_2SO_4.

Observation: The fibres turn yellowish brown first and then turn blue.

Conclusion: Blue colour indicates the presence of cellulose

2. *Test for Cutin*

Materials : *Ficus* or *Nerium* leaf, microscope, slides, coverslips, watch glass, water, KOH solution etc.

Procedure: Cut cross sections of *Nerium* leaf, mount the sections on a slide, stain them with KOH solution. Put coverglass and observe the colour of the outermost deposition on the epidermis.

Observation: The deposition takes a yellow colour.

Conclusion: Yellow colour indicates the presence of cutin.

3. *Test for Hemicellulose*

Materials: Soyabean seeds, Iodine, water, slide, cover glass, glycerine etc.

Procedure : Cut a thin section of the seed, place it in water and stain it with iodine solution for few minutes. Observe under the microscope.

Observation : Sections particularly the walls take a blue colouration.

Conclusion : Blue colouration indicates the presence of hemicellulose.

4. *Test for Suberin*

Materials : Natural cork, sudan IV (alcoholic), 50% ethyl alcohol, slide, cover glass, water etc.

Procedure : Cut thin sections of cork. Stain them with sudan IV for about 15 minutes. Wash excess stain with 50% ethyl alcohol. Transfer the sections to water for about 5 minutes and finally mount in glycerine. Put a cover glass over the section and observe under the microscope.

Observation and conclusion : All the suberized portions take red stain, indicating the presence of suberin in the cell wall.

5. *Test for Lignin*

Materials : Wood shavings, sections of woody material, 1% alcoholic phloroglucin, 25% Hcl, 1% neutral aqueous $KmNO_4$, Amonium hydroxide or sodium bicarbonate, slide, cover glass, water etc.

Procedure : (1) Treat the material (section or wood shavings), with phloroglucin on a slide and put a coverglass. Flood the section with HCL along the edges of the cover glass. Observe.

(2) Treat the material with potassium permanaganate ($KmNO_4$) solution for about 15-20 minutes. Wash the material with 25% Hcl. Follow it up with several washes of water. Add a few drops of either NH_4OH or sodium bicarbonate solution.

Observation and Conclusion : In the first method lignified walls take up a red-violet colour. In the second method a deep red colour develops.

6. *Test for Mucilage :*

Materials : Seed coat of linseed, 10% $CuSO_4$ solution, 10% KOH solution, water, slide, cover glass etc.

Procedure : Cut thin sections of the seed coat and soak them in 10% $CuSO_4$ solution for about 20 minutes. Wash them in water and transfer to 10% KOH solution. Observe the sections under microscope after mounting them in glycerine.

Observation and conclusion : Cells having mucilage are stained bright blue.

7. Test for Latex

Materials : Latex of *Calotropis*, *Asilepias*, *Euphobia* etc sucrose, Conc H_2SO_4, test tube, water etc.

Procedure : Extract the latex in alcohol (mix alcohol and latex thoroughly; centrifuge and take the supernatant) and add an equal amount of sucrose and sulphuric acid.

Observation and Conclusion : The colour turns pinkish purple. This indicates the presence of latex.

8. Test for Glucose (Fehling's test)

Materials : Glucose solution, Test tube, spirit lamp, freshly prepared solution A and B.

Procedure : Mix about 1.5 ml each of Fehyling's solution A and B in a test tube and add 3ml of glucose solution. Boil on spirit lamp. Cool it and observe.

Observation and Conclusion : A brick red pricipitate develops indicating the presence of glucose. Fehlings solution has cupric sulphate, sodium hydroxide, and potassium sodium tartarate. When mixed with glucose and boiled, the aldehyde groups of glucose are oxidized while the cupric salt gets reduced to cuprous oxide which forms a brick red pricipitate insoluble in water.

$$C_6H_{12}O_6 + 2Cu\,(oH)_2\ C_6H_{12}O_2 + Cu_2o + H_2o$$

9. Test for Glucose (Bendecit's test)

Materials : Glucose solution, Bendict's solution, spirit lamp, test tube etc.

Procedure : Add 1 ml of glucose solution to 3 ml of Benedict's solution taken in a test tube; warm the mixture on the flame of spirit lamp untill a red brown colour appears.

Observation and conclusion : The colour changes from blue to green and finally to rust brown indicating the presence of reducing sugars.

10. Test for Sucrose :

Materials : Sucrose solution, Fehling's solution A and B, NaOH solution, HCl litmus paper, spirit lamp

Proceudre : Take 3 ml of sucrose solution in a test tube and few drops of conc. HCl and boil on the flame of spirit lamp, cool. Add a little NaOH to neutrialize the mixture. Test the pH with litmus paper. Mix Fehlings solution A and B. Add 3 ml of this mixture to the neutralized mixture of sucrose solution. Boil again till a brick red precipitate appears.

Observation and Conclusion : Sucrose is a non reducing sugar, but on acid hydrolysis it gets broken up into glucose and fructose which are reducing sugars. These on treatement with Fehling's solution give a brick red precipitate.

11. Test for Starch

Materials : Starch, supspension, test tube, Iodine solution, spirit lamp.

Procedure : Take 5 ml of starch suspension in a test tube and add equal volume of idoine solution. watch the change in colour.

Observation and Conclusion : Starch gives a blue colour on reacting with iodine. The colour disappears on boiling but reappears on cooling. Starch molecules form a complex with iodine solution under normal temperature.

PROTEINS

12. Biuret Test

Materials : Gram flour, test tube, 40% NaOH, 1% CuSo4 solution.

Procedure : Mix some quantity of gram flour and prepare a solution. Take 5 ml of this solution in a test tube and add 2 ml of NaOH and one drop of $Cuso_4$ solution. Mix the contents by stirring and note the colour change.

Observation and Conclusion : A pale purple or pink colour appears indicating the presence of proteins. Biuret reagent reacts with the peptide bond of proteins and produces the colour.

13. Xanthoproteic Test

Materials : Gram flour, con. HNO_3, conc NaOH solution, spirit lamp etc.

Procedure : Mix a little gramflour with water and prepare a solution. Take about 3 ml of this solution in a test tube and add equal volume of conc. HNO_2. A thick white precipitate appears. Boil the solution, the precipitate turns canary yellow. Cool the test tube with cold water and add NaOH to neutralize and then to make it alkaline. Observe the colour.

Observation and Conclusion : Yellow colour turns to orange showing the presence of proteins. Aromatic aminoacids present in the protein when heated with HNO_3 form nitroderivative of aromatic aminoacids. These nitroderivatives on treatment with NaOH form salts which are orange coloured.

FATS

14. Test for Fat

Materials : Seeds of almond, sunflower etc., water, osmic acid, sudan III (0.5% in 70% alcohol), glycerine etc.

Procedure : Cut thin sections of the material; stain them with sudan III for about 10 minutes. Wash the sections with water. Observe the colour under microscope.

Observation and Conclusion : Materials turn red on microscopic observation. It is seen that the oil droplets have taken the red stain. In another method a few drops of osmic acid is added to the material taken in a test tube. Formation of black colour indicates the presence of fat.

15. Test for Fat (oil)

Materials : Olive oil, water, petroleum ether, sudan III (0.5% dye in 70% alcohol)

Procedure : Take a few ml of the oil in a test tube and add a few drops of sudan III. Shake well. The oil turns red. Pour half the contents into another tube. Add water to one tube and petroleum ether to another. Shake both the tubes gently and observe.

Observation and Conclusion : The red coloured oil does not mix with water hence it floats forming a red coloured ring over water. However when petroleum ether is added, oil dissolves in it and a homogenous red solution is formed.

16. Test for presence of fatty acids by saponification :

Materials : Butter, 10% alcoholic KOH, test tube, litmus paper, spirit lamp, water, NaOH solution.

Procedure : Melt the butter and take 1 ml of the oil in a test tube and add 2 ml of alcoholic KOH. Boil the mixture for about 2 minutes and add 5 ml of distilled water. Boil again for 5-6 minutes. Cool and add a few drops of conc HCl to make the mixture slighly acidic. Test the pH with litmus paper. A layer of fatty acids is formed at the top. Transfer it to another test tube and add a dilute solution of NaOH and heat.

Observation : The solution turns soapy

Conclusion : The soapy solution consists of fatty acids. By the addition of NaCl solution, the soapy solution precipitates indicating the presence of the fatty acids.

17. Test to Determine the saponification value of Fat

Materials : Butter, petroleum ether, 0.5 N alcoholic KOH, 0.5 N HCl, buretle, relex condenser, conical flask etc.

Procedure : Take one gram of butter in a conical flask (250 ml). Add 3 ml of fat solvent (petroleum ether) and shake to dissolve butter. Add 25 ml of KOH solution and attach a reflex condeser to the flask and heat the contents in a water bath for 30 minutes. Cool the mixture to the room temperature. Titrate the mixture using a burette against 0.5 NHCl. Note the reading; conduct a blank experiment (without butter) and note the reading.Find out the difference in values.

Observation and conclusion : The difference in titrated value between blank (without fat) and experiment (with fat) will be the number of mg of 0.5 N KOH required for the saponification of 1 gm of butter. This value is called the saponfication value for one gm of fat.

NUCLEIC ACIDS

(Microchemical test)

18. Test for RNA and DNA :

Materials : Onion roots, azure B solution, butanol (0.25 mg/ml in citrate buffer at pH 4), watch glass, microscope.

Procedure : With the help of a razor blade cut thin long sections of root tips of onion. Transfer them to a watch glass and stain them using Azure B solution for about 2 hrs. Wash the sections in distilled water and transfer them to another watch glass. Flood the sections with butanol for about 30 minutes. Mount the sections on a slide, put cover glass and observe under microscope.

Observation and Conclusion : Certain portions of the cells turn greenish blue and certain others turn deep blue. Greenish blue colour shows the presence of DNA and deep blue colour indicates RNA.